AFRIKA-STUDIEN Nr. 2

Publication Series "Afrika-Studien" Edited by Ifo-Institut für Wirtschaftsforschung e. V. München in Connexion with

Prof. Dr. Dr. h. c. RUDOLF STUCKEN, Erlangen
Prof. Dr. HANS WILBRANDT, Göttingen
Prof. Dr. EMIL WOERMANN, Göttingen

Editors in Chief:
Dr. phil. WILHELM MARQUARDT, München,
Afrika-Studienstelle im Ifo-Institut,
Dr. habil. HANS RUTHENBERG, Berlin,
Institut für ausländische Landwirtschaft

IFO-INSTITUT FÜR WIRTSCHAFTSFORSCHUNG
AFRIKA-STUDIENSTELLE

Agricultural Development in Tanganyika

By

HANS RUTHENBERG

With one Map and 33 Tables

SPRINGER-VERLAG BERLIN HEIDELBERG GMBH 1964

SPONSORED BY THE FRITZ THYSSEN-STIFTUNG, KÖLN

Additional material to this book can be downloaded from http://extras.springer.com

ISBN 978-3-540-03088-1 ISBN 978-3-662-30235-4 (eBook)
DOI 10.1007/978-3-662-30235-4

Ursprünglich erschienen bei Springer-Verlag OHG, Berlin · Göttingen · Heidelberg · New York 1964

Library of Congress Catalog Card Number 64—8036

Titel-Nr. 7311

Preface

In early 1961 the Ifo-Institut für Wirtschaftsforschung (Ifo-Institute for Economic Research) established an "African Studies Centre" *with the support of the Fritz-Thyssen-Foundation* to conduct research into the economic and political problems of developing countries, with special reference to the African nations. By means of investigations into the structure and prospects of development in the developing countries, the Studies Centre will aim at contributing towards the creation of a factual groundwork, on the basis of which the Federal Republic's development policy can be carried out effectively.

In this "age of development" the social and economic sciences are confronted with manifold tasks. In solving these, *interdisciplinary co-operation* will prove advantageous and even essential, if wrong and narrow judgements are to be avoided. Co-operation between scientists and institutes engaged in different branches of research will make possible a better understanding of the complex functional relationship which determine economic events and their mutations outside Western industrial society than could be normally achieved by an isolated approach. Intelligent overall economic planning, for example, requires technological analysis just as much as a precise knowledge of social structure or physical context. At this point the economist must co-operate closely with the engineer if he wishes to avoid losing his way in general patterns of growth. For, "dès lors que nous parlons (du développement) en quantités *globales*, nous ne parlons de *rien*" (Louis-Joseph Lebret). It is no less important for the economist to take into consideration the results of, for instance, sociological or geographical research.

When, in 1961, the Ifo-Institute decided to include in its research activities questions concerning developing countries and development policies, it intended not only to confine itself to purely economic considerations but to bring together all related disciplines. *Agriculture* has so far made the largest contribution to the gross domestic product of the individual economies of Africa, and in the forseeable future the success or failure of all the efforts of politicians concerned with the development of their country will basically depend on a meaningful use of agricultural resources. The first logical consequence of this for the African Studies Centre of the Ifo-

Institute was to seek support from the side of agricultural and economic research. In this desire it was supported by the *Institut für ausländische Landwirtschaft in Berlin* (Institute for Foreign Agriculture in Berlin) which was guided by similar considerations. In 1961 the two institutes agreed upon close co-operation — both in matters of personnel and research — in their work in Africa and this has since borne fruit.

In volume 2 of "Afrika-Studien" an investigation by HANS RUTHENBERG of "Agricultural Development in Tanganyika" will be presented as a first result of this co-operation. It is a description of agricultural conditions prevailing in Tanganyika and of the measures taken to encourage African peasant farming. It comprises an evaluation of experiences gained under British colonial administration and an outline of the changes brought about since the accession to independence. In 1961 the IBRD published the first comprehensive study on the economic and cultural situation in Tanganyika. The study by HANS RUTHENBERG may be regarded as supplementing that work. It focusses attention on peasant production. It indicates how progress in farming was and is encouraged in Tanganyika, which measures proved successful, which a failure, and the aims of government policy since independence. In a final chapter questions of principle in regard to agricultural development aid are discussed with special reference to Tanganyika. Thanks are due to many Africans, Europeans and Indians in Tanganyika who discussed with the author their position, their problems and their intentions and whose ideas are reflected in this publication.

Further investigations undertaken in co-operation with the Institutes for Foreign Agriculture in Berlin and Göttingen and the Institut für landwirtschaftliche Betriebs- und Arbeitslehre at Göttingen (Institute for Agricultural Methodology), have been started or are in preparation. The results of these studies will also be published in this series.

Our attempt at interdisciplinary co-operation through close working contacts with related branches of science was fortunately not limited to purely economic (in the narrower sense) and agricultural research. Thanks to the generous support received from the *Fritz-Thyssen Foundation* co-ordination of research activities in East Africa was also achieved with the *Max-Planck-Institutes* for Nutrition Physiology and Behavioural Science. Contacts were established with ethnologists, sociologists, geographers, lawyers, veterinary scientists, botanists and zoologists. Thus a research project within a regionally limited area — East Africa — came into being, aimed at throwing light on the principles of economic development and the discovery of the most efficient methods of coping with the new tasks. The establishment of a research programme leads to a better understanding of the importance of the task and of the difficulties involved. It will not be possible to cope with the problems facing us by adopting a "comprehensive approach" right from the start. Individual problems must be tackled regionally and in accordance with the facts. This may easily give the impression of a scattered collection of individual data lacking a unifying purpose. We are fully aware of this danger; but we are confident that it will be possible to blend the individual components into a single whole. Let us not forget that we are only at the beginning. After a few studies have been presented the reader will obtain a clearer picture of the ideas on which our programme is based than is possible at the moment. The institutes and scientists co-operating in this work will do their best to fill in gaps and to present a broad canvas. In the furtherance and promotion of these objectives the African Studies Centre will act as co-ordinator.

After this promising initial period of co-operation in the German scientific sphere it will also be our concern to establish close contacts with non-German, international and African scientists and institutes. To this end prospects are favourable. We are well aware of the limited scope of our national facilities for research, hence the regional limitation of co-operation between the different branches of science to the Federal Republic. Numerous research projects can only be effectively carried out on an *international* level, and others only attain their full importance if measured by universal standards.

A French institute recently attempted to ascertain the most important subjects necessary to an international research project on problems of economic and social development. In a preliminary report MICHEL DEBEAUVAIS[1] made the following statement: "Les sciences humaines doivent jouer un rôle essentiel dans l'étude du sous-développement. Elles sont actuellement *très en retard* sur les progrès des sciences exactes et des sciences naturelles, parce qu'elles répondent à des besoins moins *directement ressentis*, que le nombre des chercheurs y est encore restraint, qu'elles se trouvent à des degrés inégaux de développement et d'organisation, que les résultats de leurs recherches sont *mal connus* par les spécialistes des autres disciplines, mêmes voisines."

In our "Afrika-Studien" series we will try to ensure that at least the last sentence of the above criticism is less justified in the future — as far as our own field of activities is concerned. If the "Afrika-Studien" should develop into a wellknown source of information for research in Africa and contribute towards broadening the quite often narrow scope of the various branches of social science our efforts would be fully rewarded.

Prof. Dr. HANS WILBRANDT
Director of the Institute
for Foreign Agriculture,
Göttingen

Prof. Dr. HANS LANGELÜTKE
Chairman of the Board of
Directors of the Ifo-Institute
for Economic Research, München

[1] TIERS MONDE, Paris, Nr. 5/1961, p. 45.

Summary of recent and forthcoming publications under the African Research Programme

At the end of July 1964 the overall research programme included the following studies dealing with general and particular economic problems. In order to keep readers informed of changes, supplementary work and the progress made towards publication each volume of "Afrika-Studien" will contain a summary of the programme as a whole. Some of the studies will appear in English or French and the rest in German.

General Economic Studies

a) Tropical Africa

N. AHMAD/E. BECHER: Development Banks and Companies in Tropical Africa (being printed as volume 1).

R. GÜSTEN/H. HELMSCHROTT: National Accounting Systems in Tropical Africa (being printed as volume 3).

N. AHMAD/E. BECHER/E. HARDER: Economic Planning and Development Plans in Tropical Africa (nearly ready).

b) East Africa

L. SCHNITTGER: Tax Systems and Fiscal Policy as an Instrument of Economic Development in East Africa (nearly ready).

R. GÜSTEN: On Problems in Connection with Economic Union in East Africa (in preparation).

R. VENTE: Methods and Results of Economic Planning in East Africa (in preparation).

F. GOLL: Israeli Aid to Developing Countries with Special Reference to East Africa (in preparation).

Agricultural Studies

a) Tropical Africa

A. REITHINGER: Possibilities of Diversifying Agricultural Production in Tropical Africa (nearly ready).

(Various): The Effects of the European Common Market Regulation of Agricultural Products on the Export Trade of Developing Countries (in preparation).

H. PÖSSINGER: Results and Problems of Agricultural Development in Portuguese Africa (ready).

b) East Africa

1. Comprehensive Basic Studies

H. RUTHENBERG: Agricultural Development in Tanganyika (appearing as volume 2).

H. RUTHENBERG: Peasant Production in Kenya and Measures to Promote it (in preparation).

2. Botanical, Breeding and Economic Aspects of Cattle Farming in East Africa

H. LEIPPERT: Natural Planting Associations in the Arid Areas of East Africa (in preparation).

K. MENN: Meat Production in the Arid Areas of East Africa (in preparation).

N. NEWIGER: Communal Forms of Animal Husbandry (and Soil Husbandry) in East Africa (in preparation).

E. RADDATZ: The Organization of African Peasant Farms and Dairy Farming in Kenya (in preparation).

B. ENGEL: The Organization of Meat and Milk Markets in East Africa (in preparation).

3. The Organization of Peasant Farming Systems in East Africa

D. v. ROTENHAN: The Organization of Land Use in Sukumaland (cotton) (nearly ready).

H. PÖSSINGER: The Possibilities and Limitations of Peasant Sisal in East Africa (in preparation).

S. GROENEVELD: The Organization of Cattle/Coconut Palm Farms near Tanga (in preparation).

W. SCHEFFLER/A. v. GAGERN: Managerial and Sociological Problems of Peasant Tobacco Production in Tanganyika (in preparation).

K. FRIEDRICH/H. JÜRGENS: The Organization of Land Use and Animal Husbandry in the Coffee-growing Area near Bukoba, Tanganyika (in preparation).

E. BAUM: The Organization of Farm and Household among the Coffee-Banana-Milk Peasants on Mt. Kilimanjaro (in preparation).

4. Other Investigations in Connection with Agricultural Development

H. FLIEDNER: The Economic and Social Impact of Land Tenure Reform in Kenya (ready).

M. Paulus: The Role of Co-operatives in the Economic Development of East Africa, and especially of Tanganyika (nearly ready).

N. N.: Nutritional Habits and Food Shortages in Northern Tanganyika (in preparation).

F. Dieterlen/P. Kunkel: Tropical Rodents and Birds as Agricultural Parasites (in preparation).

W. Kühne: A Study of Animal Behaviour in the Serengeti (in preparation).

Studies in Commerce and Trade

H. Kainzbauer: Trade in the Economic Development of Tanganyika (in preparation).

K. Schädler: Handicrafts in the Economic Development of Tanganyika (in preparation).

Sociological Studies

A. v. Molnos: Methods and Results of Sociological Research in East Africa (nearly ready).

H. Harlander/A. v. Molnos: The Role of Women in the Economic and Social Development of East Africa (in preparation).

O. Raum: The African's Willingness and Ability to Adapt himself to the Modern Economy, with special reference to the Kilombero Valley, Tanganyika (in preparation).

N. N.: The African as a worker in Industry in East Africa (in preparation).

Various Regional Studies

W. Marquardt: The Interrelations between Man, Nature and Economic Organization with special reference to Madagascar (in preparation).

R. Güsten: Problems of Economic Development as instanced in the Sudan (ready).

O.-O. Neuhoff: The Economics of Raw Material in the Development Planning of the Republic of Gabon (nearly ready).

H. Jürgens: Contributions to Domestic Migration and Population Development in Liberia (appearing as volume 4).

H. D. Ludwig: Ukara — a Development Study in Economic Geography (in preparation).

K. Schädler/N. N.: Development Possibilities in Ulanga-District, Tanganyika (in preparation).

Bibliographies

D. Mezger/E. Littich: Recent English and American Economic Research in East Africa. A Selected Bibliography (in preparation).

Contents

List of Tables

Chapter A

The Role of Agriculture in the Economic Development of Tanganyika

I. Land and Population

Tanganyika, the largest and most populous country within the East African Common Market, covers an area of 341,150 sq. miles — four times that of the Federal Republic of Germany. In 1961 the population was estimated at 9.4 million, including 22,700 Europeans, 90,500 Indians, Pakistanis and Goans, and 39,000 Arabs. Though numerically insignificant, the minority groups play an important part in the economic and social life of the country.

Tanganyika is not a natural geographic unit. To a large extent its borders are a legacy of the political situation towards the end of the 19th century. Following the natural features of the country, the population is concentrated along the periphery: on the coast, in the rainy highlands of the South, around Mt. Kilimanjaro and Mt. Meru, and along the shores of Lake Victoria. The thinly populated central part of the country is covered with bush and steppe. However, the present borders have existed for several generations; there is a central railway; there are administrative and political differences distinguishing Tanganyika from its neighbours; the use of Swahili and English is exercising a unifying influence; and, in particular, the way political ideas are manifesting themselves nowadays — all this combined in an originally artificial colonial entity gives it so much in common that one can indeed speak of a nation, and of NYERERE as 'Father of the Nation' [1].

Nevertheless, there are long distances between the various settlements and consequently strong differences between the African population groups. The relatively prosperous Wachagga around Mt. Kilimanjaro have little in common with the emphatically traditional cattle-herding Massai. The problems of the Wasukuma who grow cotton near Lake Victoria, are quite

[1] J. K. NYERERE, President since 1962, is the leading political figure in Tanganyika.

different from those of the Wagogo who live in the dry interior of the country and are constantly threatened with famine. In all, there are about 120 tribes in Tanganyika. They have different traditions and customs, varying access to markets and schools, and differ in their economic behaviour. Common to all of them is that they are peasants and herdsmen who still live in traditional village and tribal households based on the principle of self-sufficiency, even though they are in contact almost everywhere with commercial life, producing agricultural products for sale and buying consumer goods.

II. The Position of Agriculture in the Economy

The Europeans and Asians — apart from civil servants — derive their livelihood mainly from agriculture, i.e. from the estate economy, from trade, the processing and distribution of agricultural products and from supplying the means of production. Hence agriculture is the dominant sector of the economy. In 1961 there were only 172.000 Africans employed in administration, trade, industry and domestic service. With the exception of 179.000 agricultural wage earners (including those in forestry and fisheries), most of the Africans are independent peasants and herdsmen.

The dominant position of agriculture is shown in the estimates of the national product. These figures need, of course, cautious interpreting. The collection and evaluation of production figures in the subsistence sector are

Table 1. *The Gross National Product of Tanganyika* [1]

	1955	1956	1957	1958	1959	1960	1961	1962
Total in mill. £ . . .	146.7	152.4	162.4	167.1	177.1	186.2	188.7	203.3
per capita in £ . . .	17.4	17.7	18.5	18.7	19.4	20.0	19.9	.
Agriculture [2] in mill. £	85.1	87.4	91.1	90.1	96.7	100.9	97.3	109.2
Agriculture as per cent of total . .	58.2	57.5	56.0	53.9	54.6	54.0	51.4	53.0
Subsistence as per cent of total agriculture. . . .	60.8	56.7	60.3	60.4	57.3	55.2	59.0	59.0

[1] at factor cost
[2] forestry not included

Sources: a) Tanganyika: Budget Survey 1962—63, Table 1—3, p. 4.
b) Colonial Office: Report for 1960, Part II, Appendix I, p. 6.
c) UN: Ninth Report of the Committee on Rural Economic Development, Annex II, p. 40.

based on "intelligent assumption" rather than on systematic enquiry. They are of value in so far as they convey some idea of the relative importance of this sector. The national product for 1962 is estimated at £ 203.3 million or roughly £ 20 per head of population, £ 109,2 million or 53 per cent coming from agriculture. Only 41 per cent of the gross national product in agriculture is brought into circulation, the remaining 59 per cent being consumed by the producers themselves. The emphasis in agriculture is, therefore, on production for direct consumption (see Table 1).

Next to direct consumption come agricultural exports. The internal market for agricultural products is by comparison of little importance. Agricultural products, in particular sisal, cotton, coffee, tea and cashew nuts account for 75 per cent of the exports to countries outside the East African Common Market. Mining and manufacture are making rapid progress. The export value of diamonds has increased during the past five

Table 2. *Factories Processing Agricultural Products*

	Number of Factories	Number employed
Sisal processing	236	29,767
Flour milling	1,258	6,734
Cotton ginning	43	4,737
Tobacco processing	54	1,891
Coffee pulping	135	1,585
Rice mills	104	1,458
Food canning	8	1,411
Oil mills	92	1,320
Tea manufacture	19	1,007
Coffee curing	11	689
Hides, skins processing	40	512
Sugar manufacture	7	496
Nuts, processing	5	446
Coconut fibre, processing	8	258
Kapok ginning	13	195
Pyrethrum drying	33	140
Milk pasteurizing	7	130
Dairy produce	7	94
Copra drying	6	63
Bacon manufacture	3	28
Papain manufacture	3	22
Honey processing	1	21
Processing of other agricultural products	13	941
Total of Processing Plants	2,106	53,845
Total Industry of Tanganyika	4,917	83,473
Percentages of processing from total industry	43	65

Sources: a) Little, A. D. (Inc.) "Tanganyika Industrial Development". Dar es Salaam, December, 1961.
b) Statistical Abstracts, 1961, p. 79 ff.

years from £ 3.2 million to £ 5.8 million, i.e. to 12 per cent of all exports. Other industrial projects are either being planned or are in course of construction. However, the non-agricultural sector of the economy is still so small that even a rapid expansion would have little effect on the dominant position of agriculture, the more so since most of the existing industries are engaged in the processing of agricultural products. This applies to 2,106 enterprises with 53,315 workers, out of a total of 4,916 industrial enterprises employing 82,953 workers (see Table 2).

Though there are mineral resources, such as coal and iron, their exploitation apparently encounters such great difficulties in transportation and markets, that for the present it is not under consideration. The "Little Report" suggests that the possibilities for further industrialization are seen mainly in an increase of the agricultural processing industries.

Consequently, the economic growth of Tanganyika is for the foreseeable future closely tied up with an increase in agricultural production and productivity. This is not meant to accord a fundamental priority to agricultural over industrial development. In every country, Tanganyika not excepted, economic development necessitates urbanization and industrialization. This implies a higher rate of growth for cities and industries rather than for agriculture. Development limited mainly to the agricultural sector usually leads to little more than "stagnation at a somewhat higher level". However, in considering the desired objectives we must not overlook the conditions prevailing at the point of departure. In a developing country lacking extensive mining or timber industries the development of other sectors of the economy depends largely on how rapidly and cheaply the existing agricultural potential is developed, especially since this potential can best be exploited through increased efforts by the country itself.

III. Objectives of an Agricultural Development Policy

1. Meeting the Home Demand

Agriculture in Tanganyika has to solve a number of problems if economic growth is to take place. The growing internal demand of farming families and local markets will have to be met. From 1948 to 1957 the annual rate of increase of the population was about 1.75 per cent. At present it is generally assumed to have reached approximately 2 per cent. Hence, even under conditions of unchanged per capita income, i.e. under conditions of economic stagnation, the demand for agricultural products would grow at the same annual rate. In addition, growing per capita incomes are accompanied by increased demand. From 1955 to 1960 the per capita national product increased by 3 per cent per annum[1]. The income elasticity for agricultural products is certainly still high. If it is set on an over-all basis at 0.7, and if it is assumed that the population will

[1] No account is taken of the relatively small price increases.

continue to grow at an annual rate of 2 per cent and the per capita national product at 3 per cent, then agricultural production will have to increase at an annual rate of 4.1 per cent in order to keep pace with the national demand. It will not be easy to attain such a rate of growth. Germany's annual rate of growth from 1881 to 1931 was about 1.8 per cent. Japan's agricultural production, known to have risen particularly rapidly, had a rate of increase of 2 per cent in the years from 1880 to 1920. Such rates of increase are not sufficient for Tanganyika. Agriculture faces the task of meeting the needs of an internal market expanding approximately twice as rapidly.

The alternative, i.e. the import of foodstuffs, raises the question of payments. It is, of course, possible for Tanganyika to obtain maize and powdered milk from the United States at low cost or even free of charge. It may be that Tanganyika can gain a comparative-cost advantage by pushing up the production of sisal, cashew nuts, cotton, and tea, and buying maize and rice. Yet in view of the existing potential in terms of manpower and land it is probably more advantageous to meet the additional need for foodstuffs by increasing home production, and at the same time to stimulate initiative and willing self-help, dormant in the villages. Since land and labour are in sufficient supply, the expanded production of sisal, cashew nuts, and tea would hardly have to compete with that of maize and rice.

2. Export

Except for the limited stimulus of the home market, Tanganyika's economy depends on agricultural exports. The building of cities and industries requires imported means of production. What Tanganyika can offer as payment for these imports are, in the first place, her agricultural products. The nation's balance of trade is not problematic. In 1961 Tanganyika's exports to countries outside the East African Common Marked exceeded her imports by £ 10 million or 25 per cent. And this although 1961 was emphatically a bad year. The expansion of agricultural exports should be seen rather as an important contribution to the East African Common Market, upon whose success progress in Tanganyika greatly depends. Moreover, foreign trade duties, especially import duties, provide the basis of the tax revenues. The possibility of financing future economic development from inland sources depends in large part upon increased foreign trade and the consequent increase in tax revenues.

3. Increased Productivity

The third major task is to increase per capita productivity. This is fundamental to any higher earnings. Peasant families would be able to buy more consumer and capital goods. The home manufacturer would receive encouragment and the Treasury an additional income from duties and taxes.

To achieve increased agricultural productivity is no simple task under the conditions prevailing in Tanganyika. It is not a question of raising the per capita production of a stable or decreasing agricultural population, as in industrialized countries. The problem is that the number of people in rural areas is increasing so rapidly that the growing non-agricultural economy is unable to absorb the total "additional" labour force. Agriculture, therefore, faces the task of incorporating productively that part of the additional manpower which cannot be gainfully employed in towns. The problem is, therefore, one of increasing the productivity of a growing rural population, and this without the extensive use of expensive capital equipment.

4. Agriculture's Contribution to the Formation of Capital

In addition to all of this, Tanganyika's agriculture will have to contribute to the formation of capital in town and industry. Capital formation means, generally speaking, that part of the production is invested in workshops, tools, equipment, roads, improvements, etc. In a broader sense it also means investments in skills, knowledge, experience, etc., though such investments can hardly be measured quantitatively.

Economic development requires large amounts of capital, and the question of where it is to come from is, therefore, of primary importance. In Tanganyika the profits from trade and industry are not sufficient to provide this capital, simply because there is too little of it. Development aid as a start, and for the purpose of overcoming certain bottlenecks, may prove very useful. But because of the number of underdeveloped countries and the extent of their investment requirements, Tanganyika must be guided by the principle that "capital is made at home" [1]. In this connection it must be remembered that a country's economic independence depends in the long run on investments being financed from the efforts of its own people.

In developing countries there is hardly any other internal source of money for capital formation than mobilization of the production potential of natural resources and re-investment of a part of the increase in productivity. Since Tanganyika is not very well endowed with forests or minerals, the country has no other choice than to finance capital formation from agriculture. However, it is not enough for agriculture to meet only its own capital needs. The key to achieving the prerequisites for economic development lies in the transfer of part of the returns of agri-

[1] Nurkse, E.: Problems of Capital Formation in Underdeveloped Countries. Oxford 1955, p. 33.

culture to towns and industries, in order to broaden the scope of employment and to increase buying power in the towns, which is in the interest of the rural population itself. Thus Tanganyika's agrarian policy faces a different and more difficult task than that of the industrialized countries. In these countries agriculture, which is relatively less important, is usually subsidized. In Tanganyika, agriculture is the dominant sector of the economy and, therefore, of necessity the "foster mother" of town and industry.

5. Conclusions

The accomplishment of the above-mentioned tasks presupposes that a way can be found to bring about a constant agricultural production increase "cheaply". The chances of achieving this lie in the fact that potential in terms of labour, land and water is not utilized very efficiently. The introduction of "new combinations", through institutional, administrative, and technical innovations, can effect considerably higher returns without the necessity of large capital outlay. The realization of these possibilities will depend upon the extent to which the initiative of the rural population can be brought into play.

The present study proposes to outline:

1) How agricultural policy was carried out under the British mandate,
2) How it has been carried out since independence,
3) Which aid measures seem most likely to contribute to successful development in agriculture, taking into account the existing conditions.

The following sections on climate, soil, markets, organization of farming, volume of production and export are limited to such information as is essential to an understanding of agricultural development policy. More detailed information can be obtained from the references quoted.

Chapter B

A Brief Survey of the Agricultural Situation in Tanganyika

I. Climate, Soil and Vegetation

1. Rainfall

For the most part Tanganyika has a dry climate, heavier rains occurring in the northern and southern highlands and some local pockets. About one third of the country receives enough rainfall to support agriculture fairly adequately (95 per cent, or greater, rate of probability of 30 inches rainfall). Another third receives sufficient rain for bush and grass only (less than

85 per cent probability of 20 inches from October to May). The remaining third of the country falls somewhere between these two. It is considered marginal land, i.e. cultivation is possible, but the average yields are low and vary greatly from season to season[1].

The rainy season is usually from October/November to May. In some parts of the country a distinction is made between the "short rains" of November and December and the "long rains" from March to May. In the highland regions rainfall is higher and more evenly distributed.

On the whole Tanganyika's rainfall is not especially low. Those regions which receive 30 inches are quite large in relation to the population. The rest of the country receives enough rainfall to support grass and bush vegetation, which indicates that so far as plant growth is concerned, cattle ranching is possible. The fact that the rainfall is unevenly distributed and falls with great intensity, so that much of it cannot be absorbed by the soil quickly enough and erosion is therefore likely, is characteristic of tropical climates in general, not only of Tanganyika. The more pressing problem here is that the rainfalls are unreliable. They fluctuate greatly from year to year and from month to month. There is also no guarantee that the rains, once begun, will continue. Farming is therefore faced with unusual hazards.

2. Soil and Water

Available information on the condition of Tanganyika's soils would seem to indicate that a large, if not the major, part is not fertile, and that extensive regions will show satisfactory yields only with good husbandry. High yields can be achieved in valley soils, for example in the "mbuga" heavy dark moist soils, but the cultivation of such soils requires extensive drainage and tractive power and is, therefore, difficult for peasant farmers. The opening up of potentially fertile alluvial river valleys means great expenditure in terms of land reclamation, flood control, irrigation, and desalinization. The only soils which may be considered fertile and easily accessible are the relatively small areas of volcanic origin in the highlands of the north and south. Where rainfall is sufficient these areas are heavily populated.

The scope for irrigation is probably very extensive. Currently irrigated soil is estimated at about 340,000 acres. The Report of the World Bank

---

[1] According to a study made by Gillman in 1934 Tanganyika can be divided into four groups with regard to water supply. 1) Good — about 10 per cent of the territory, in which 63 per cent of the population is concentrated; 2) sufficient — 8 per cent of the territory, with 18 per cent of the population; 3) poor or sporadic — 20 per cent of the territory, inhabitated by 18 per cent of the population; 4) practically no water available — 62 per cent of the territory, which is practically uninhabited. (See Gillman, C.: *The Population of Tanganyika*, p. 28 ff.)

indicates that about 4 million acres could be made arable through irrigation and flood control. At the moment hardly anyone is able to assess accurately the potential in terms of water and soil of the vast, almost uninhabited, regions in the south and southwest of the country.

Statistics classify 26.9 per cent of the land area as suitable for cultivation — a high percentage in relation to the size of the country (see Table 3).

Table 3. *Land Distribution*

Total Land Area, excluding the Lakes: 341,150 sqm.

Land Cover		Land Tenure		Land Use	
in percentages of total area					
Vegetation actively induced by man	9.0	Cultivated by Africans	7.7	Arable land, fallows, temporary meadows	8.8
Grassland and wooded grassland	36.0	Alienated	1.1	Permanent meadows and pastures	9.9
Miombo forests	35.0	Forest Reserves	9.8	Productive but uncultivated land	26.9
Woodland — bushland intermediate	16.0	Game Reserves	7.7	Wood or forest land	36.4
Closed forest and forest woodland intermediate	2.0	Not specified	73.7	All other land	18.0
Swamps	1.0				
Desert and semi-desert	1.0				
Total	100.0		100.0		100.0

Note: These figures are calculated estimates. They vary greatly from year to year even in different issues of the same publication.

Sources: a) IBRD, The Economic Development of Tanganyika, Table 15, p. 48 and Table 59, p. 259.
b) Colonial Office: Tanganyika. Report for the year 1960, Appendix VIII.

These figures must be used carefully. The classification "suitable for cultivation" is drawn rather arbitrarily, being based on somewhat obscure technical farming criteria.

Tanganyika is, therefore, not suffering so much from a shortage of soil and water. The difficulties are rather: 1) that land reclamation is expensive in relation to the monetary returns which can be expected; 2) that the use of small fertile pockets situated amid vast unfertile areas is costly; 3) that the peasants have neither the knowledge nor the means to utilize potentially good land; 4) that the people do not live where the soil is good; 5) that they are not inclined to move to such areas; and 6) that traditional attitudes lead to the peasants being content with farming relatively small plots of land.

3. Vegetation

Due to soil and climatic conditions, 87 per cent of the land area is covered by a rather thick tree-bush-grass steppe. Table 3 indicates that 16 per cent is covered by thick bush, 35 per cent by a mixture of trees (Miombo), bush, and grass, and 36 per cent by grassland and wooded grassland. Bush vegetation is a costly barrier to agricultural development. The cost of clearing this land is high in relation to the poor and uncertain yields. Fully-mechanized land clearing as done under the Groundnut Scheme amounted 300 shs/acre; half-mechanized costs 156—176 shs/acre [1]. The costs of land reclamation in the fertile Kilombero Valley (Kiberege area) are estimated at between 1000 shs and 1500 shs per acre.

On the other hand, the returns on such products as maize, sorghum, groundnuts, etc., which might be expected under normal circumstances, fluctuate between 100—300 shs/acre. The cost of bush clearance corresponds to the value of at least one season's harvest, which is high for marginal land where returns hardly cover the usual costs of planting, cultivation and harvesting. Figures of monetary costs do not apply to peasants using hand inplements and family labour. They do, however, illustrate that the peasant has to be satisfied with very small, uncertain returns as soon as he clears further marginal bush land. Land clearance is not the only problem. The soil is rarely fertile enough to support continuous farming. During the inevitable longer periods when the land is kept fallow the bush re-asserts itself and additional efforts are needed for re-clearing.

Another consequence of bush vegetation is the spread of the tsetse-fly. 60 per cent of the land area is infested with this pest. Animal husbandry is possible only with constant and careful veterinary supervision. It is technically possible to suppress the tsetse-fly by selective bush-clearing, but this method is costly, considering the low yields per acre in ranching.

Tanganyika is, therefore, in an economically unfavourable position, with climates that are neither dry nor wet enough. If there were less rainfall the vegetation would more resemble that of grassland. Farming would not be burdened with such high costs for land clearance; the tsetse-fly could be more easily controlled; animal husbandry or the use of ox-drawn ploughs would be possible, irrigation would be made simpler and less expensive; and destruction due to floods and unexpected rains less frequent. On the other hand, more plentiful rainfall would result in higher and more certain crop yields.

[1] Overseas Food Corporation, Reports and Accounts, 1954—55, p. 52.

II. Prices and Markets

A decisive obstacle to the development of the considerable, but marginal, potential lies in the unfavourable price situation. The internal market, which has little buying power and is split into numerous small widely separated markets, has an insignificant turnover. The 1957 census indicates that only 308,000 people from a population of at that time 8.7 million were living in towns of over 4,800 inhabitants. The per capita buying power of wage earners is low. In 1960 the monthly income of African male employees averaged 92 shs. Since then, however, conditions have changed considerably. The urban population has grown and wages have greatly increased, especially in 1962. In the fall of 1962 the minimum wage for Dar-es-Salaam was set at 150 shs. Useful as such incentives may be for stimulating the internal demand for agricultural products, they hardly affect the dominant position of exports and of subsistence production.

The limited scope and scattered nature of the internal market and the long, difficult lines of transportation combine to keep crop prices low. In 1960, maize was sold on local markets at a national average of £ 12 per ton. Prices fluctuate greatly from year to year and from place to place. Maize, for example, varies between £ 10 and £ 30 per ton. In some years Tanganyika attains a surplus of maize for export, in others it has to import maize. The home consumption of such products as coffee, cashew nuts, sisal etc., is negligible and will probably rise only gradually.

The farmer has no other choice, therefore, than to concentrate on raising such products as can be exported. Consequently the prices for agricultural products in Tanganyika are subject to world market conditions, which in turn have worsened since the Korea boom. The f.o.b. price for coffee per ton in 1954 was £ 515, in 1961 only £ 275. In the case of cotton it was £ 423 in 1952 and £ 229 in 1962. The price for sisal per ton fell from £ 166 in 1951 to £ 52 in 1957—1958; in 1962, however, it rose to more than £ 100. Tea is the only crop which experienced a strong and constant rise in price (see Table 4).

Cattle prices have developed rather favourably. In 1951 the price was £ 7 per head for a liveweight of 500 lbs. In 1960 it was £ 9. Milk prices fluctuate greatly according to sales. In areas with relatively prosperous consumers, i.e. near towns, in the coffee-growing regions of the Wachagga, or near sisal plantations, 3—4 shs/gal. is paid. In more isolated regions where milk is used for ghee-making, it is priced at about 1 sh/gal.

Low prices are especially evident in the more inaccessible areas. It pays to produce bulk crops for the limited local demand only. Away from the coastal towns, i.e. off the few main roads which are passable all the year

Table 4. *Prices for Principal Crops*
£/t

	1951	1952	1953	1954	1955	1956	1957	1958	1959	1960	1961	1962
Average Export price f.o.b.												
sisal	166	137	75	65	57	58	52	52	63	75	70	72
cotton lint	334	423	324	281	272	268	242	226	217	227	229	227
coffee	266	296	378	515	373	427	386	341	293	292	275	256
cashew nuts	41	46	42	34	48	53	45	35	47	58	45	·
tea	293	303	302	315	317	274	276	270	287	364	423	·
Price to producers at local markets												
maize	·	·	·	21	18	16	14	15	14	12	19	19
rice (paddy)	·	·	·	30	29	30	30	35	24	21	26	24
groundnuts	·	·	45	54	50	50	47	48	47	49	·	·
castor seed	·	·	·	30	·	47	47	30	38	40	·	·
sesame	·	·	·	49	·	50	52	60	48	47	47	51
cassava (dried)	·	·	·	9	·	10	12	11	11	8	12	14
millet	·	·	·	·	17	18	16	17	20	17	21	23
sorghum	·	·	·	·	·	·	18	18	15	13	16	20
mixed beans	·	·	·	·	24	24	24	24	25	22	26	31
pigeon peas	·	·	·	·	19	18	20	25	19	20	30	26
sunflower seed	·	·	·	·	20	19	18	14	16	14	16	16
cashew nuts	·	·	·	·	42	45	33	26	30	35	·	·

Sources: a) Annual Reports, Department of Agriculture.
b) Quarterly Economic and Statistical Bulletin, East African Common Services Organization (High Commission).
c) Statistical Abstracts 1961, Table 6, p. 65.

round, the only profitable crops are those which can be transported cheaply because of their high value per unit of weight (tea, coffee, cotton, pyrethrum and possibly oilseed).

III. Organization of Farming

Agricultural activities which are carried out under the aforementioned natural and economic conditions may be divided into three groups which have little in common:

the estate economy of Europeans and Asians,
crop cultivation of African peasants,
animal husbandry of peasants and herdsmen.

1. The Estate Economy

Agricultural estates and plantations cover about 2.5 million acres or 1.1 per cent of the land area. Only about one third of this area is used for plantation crops and other farming. Another third is used for grazing. The remainder consists of fallow land, forest and unspecified areas. The land occupied has been for the most part excluded from African use since the period of German colonization (see Tables 5 and 6). The possession and management of estates is almost exclusively in the hands of Europeans and Asians. 60 per cent are owned by British citizens, including those from South Africa. 14 per cent belong to Greeks. Indians and Pakistanis together possess 10 per cent of this land. Some of the Europeans and Asians have since taken Tanganyikan nationality.

Although in some respects out of date, particularly with regard to the employment figures, the census of 1958 offers the following data:

Sisal: As far as area and economic importance is concerned, sisal cultivation occupies first place. According to the census of 1958, sisal is grown on 536,296 acres, for the most part near the coast and along the two railway lines. 62 of the 151 estates are in the Tanga Region, 55 in the Eastern Region. These estates have an average size of 3,500 acres, employ about 700 African farm workers and produce 1,400 tons of fibre annually. The following special features of sisal are noteworthy:

— The establishment of a sisal plantation requires very high initial investments for land clearance, planting, water supply, factory, and transportation of leaf. According to GUILLEBAUD, the gross investment necessary to produce 1 ton of fibre annually, i. e. on approximately 2.6 acres, was eight years ago at least £ 200.

— Sisal production entails high expenditure on wages. Inquiries made in 1956 indicate that the wages bill (without salaries) constitutes 38 per cent of the total costs. The recent wage increases will raise this percentage even more.

— Sisal production must be planned for years in advance. The growth cycle is about ten years. The significant advantage of sisal is that it is a simple, robust, adaptable and highly resistant plant.

Coffee: Of a total of 132 coffee estates covering 19,009 acres, 70 are in the Northern Region, on the slopes of Mt. Kilimanjaro and Mt. Meru, and near Oldeani. 36 estates are in the Southern Highlands. The 1958 census indicates that

Table 5. *Land in the Estate Economy* 1958 (in 1,000 acres)

Estate Crops	
sisal	536,3
coffee	19,0
tea	14,7
sugar cane	13,3
kapok	7,7
coconut palm	6,9
cashew nuts	6,2
papain	3,7
pyrethrum	2,5
wattle	0,9
rubber	0,3
Total	611,5
Other Crops	
maize	41,9
seed beans	11,8
wheat	6,4
tobacco	5,6
soya beans	4,4
sunflower seed	4,1
groundnuts	3,9
sorghum	2,8
barley	2,5
castor seed	2,1
rice	0,8
millet	0,3
potatoes	0,2
cassava	0,1
oats	0,1
fodder plants	5,1
other	8,1
Total	100,2
Remaining Land	
grazing, fenced in	38,2
grazing, not fenced in	646,5
hay	4,2
fallows	102,3
cultivated by workers	17,3
forest	225,2
swamps	162,8
not specified	238,5
Total	1435,0
All totals	2146,7

Source: Tanganyika Agricultural Census 1958, Table 19.

the average size of these estates is 144 acres. Numerous estates raise papain, maize, etc. in addition to coffee. Each of these estates employs an average of 73 African workers.

Table 6. *Long Term Rights of Occupancy*
as of 31st December 1960

Nationality[1]	Number of Holdings	Acreage (in 1,000 acres)
British (excl. South Africans)	470	1,316
Greeks	279	355
Indians and Pakistanis	287	256
British (from South Africa)	107	192
Dutch	14	33
Germans	45	32
Swiss	24	31
Danes	11	31
Africans	35	17
Missions	265	17
Arabs	34	10
Goans	11	6
Italians	7	5
Syrians	4	3
French	4	3
Americans (USA)	3	1
Other	66	181
Total	1,666	2,489

[1] Nationality before the independence of Tanganyika. Since then some of the persons included in these statistics have become citizens of Tanganyika.

Source: Colonial Office: Tanganyika. Report for the Year 1960, Part II, p. 51.

Tea: In contrast to coffee which is grown principally on smaller estates, tea is produced on 16 estates in the Southern Highlands and on 9 estates in the Usumbara Mts. (total acreage 14,719). The estates average 890 acres for the cultivation of tea, and produce an average of 100 tons each of dried tea valued at £ 31,000. It should be noted that the tea industry was rapidly expanding at the time of the census. A very large part of the area used for production included young, not yet fully matured plants.

Other Types of Farming: The 1958 census counted 30 sugar cane plantations with a total of 13,303 acres. Of special importance are the large sugar cane plantations at Arusha, Bukoba and the Ruaha. The last two are new and not included in the census. The Eastern Region is the principal location for sugar cane production.

Tobacco is cultivated on 50 estates in the Southern Highlands[1]. The production of wheat, pyrethrum and seed beans is principally important in the Northern Region There are also some ranches, notably the large one in Kongwa, which is run by the Tanganyika Agricultural Corporation, the successor to the groundnut project.

Although the estates possess only 1.1 per cent of the total land area and 10 per cent of the arable land — but frequently that of higher grade soil —

[1] Some of these estates stopped production in 1962.

they are of great economic significance. According to an estimate made in 1954, they account for 25 per cent of the total agricultural product. The Report of the World Bank indicates that 35 per cent of the market supply

Table 7. *Type of Farm by Predominant Activity* (1958)

Type	Number of Farms
cereals	155
sisal	151
coffee	132
ranches	106
mixed farming . .	105
tobacco	60
oilseeds	52
sugar	30
tea	25
pyrethrum	21
seed beans	19
kapok	19
papain	10
wattle	7
rubber	1
cinchona	1
not specified . . .	4
Total	898

Note: According to the census, these 898 farms constitute 75—85 per cent of the non-african farms. The remaining 15—25 per cent consist of relatively unimportant small farms.

Source: Agricultural Census 1958, Table 20.

and 45 per cent of the agricultural exports are provided by these estates. In the last 10 years there has been considerable expansion in sisal and tea. Of late there has been a strong production increase on coffee plantations, due to the joint action of plant protection measures and mineral fertilizers. Technical advances can be applied more effectively to the intensively farmed coffee monoculture of the estates than to the comparatively extensive mixed coffee and banana cultures of the African coffee farmers.

The importance of the estate economy is not limited to production and export. It has introduced new products and opened up markets to which the African farmer on his own would hardly have gained access. In many respects the estates have created the conditions necessary for agricultural development; they introduced new ideas, organized the marketing and processing, maintained repair shops, gave contract work to farmers, trained workers, etc. No less important are the effects in other sectors of the economy. Transport and processing industries in Tanganyika depend much more on agricultural estates where division of labour is practiced than on African farms working on a semi-subsistence level. The estates are the major employers. Of a total of 400,000 African workers in 1960, 199,000 were employed in agriculture, i.e. mostly on agricultural estates. 38 per cent of all wages earned by Africans were paid by them. Their contribution to the tax revenues is of great importance.

This does not mean that all the estates are well run. The acquisition of land in Tanganyika has frequently been motivated by speculative considerations. After two world wars requisitioned German holdings were relatively cheap. It was not only the experienced farmer who took advantage of this. Some estates are so small that one can hardly speak of large-scale production. In many instances only a part of the arable land is put to use. The large discrepancy between well farmed and badly run estates indicates here, as it does all over the world, that even at the present level of cultivation, production could be higher and more profitable. Nevertheless, the average quality of land use on agricultural estates is relatively good, i.e. undoubtedly better than the average of the African farmer.

The expansion of the estate economy would probably be an economically simple and fruitful starting point towards increased agricultural production and productivity in Tanganyika. In view of the social and political situation, especially as far as European and Asian owners are concerned, this is something to be thought of in exceptional cases only. Among the Africans, the idea is growing that land as a factor of production is in short supply. They mistrust the leasing of long-term rights of occupancy because their own extensive use of land leads them to believe that there is a shortage. It is in Tanganyika's interest to maintain a favourable economic climate for the estate economy, the guaranteeing of acquired rights of occupancy being of paramount importance. An increase in the number or size of European and Asian possessions would merely endanger the present fairly realistic attitude towards the existing estates. Hence, the only way for agriculture to develop is to concentrate efforts on the peasant sector.

2. Peasant Farming

Commercial peasant production originated in the work of the missions. The planting of cotton in the Lake Region (Ukerewe) and of coffee around Mt. Kilimanjaro was started by Catholic missionaries. During the last decade of the German administration, peasant production was systematically furthered and supported by extensive investments in research. Under British administration, and especially in the years 1950 — 1960, it made rapid progress. During this time a small number of well-trained agriculturalists in colonial service, assisted materially and financially by development aid from Britain and encouraged by the results achieved in Kenya and Rhodesia, worked with visible success towards improving the output of African farms.

In view of the fact that there are about 400,000 coffee farmers, 250—300,000 cotton farmers and about 60—80,000 pyrethrum farmers, half the farmers in Tanganyika are tied to the commercial production of these three products alone. According to estimates made in 1954, 75 per cent of the total agricultural product, 65 per cent of the produce offered for sale and 55 per cent of the exports, came from African farms. Cotton, millet, sorghum, root crops, bananas, leguminous crops, rice, etc., are produced almost entirely on African farms, which also produce approximately one-half of the country's coffee and tobacco (see Table 8). In 1954, pyrethrum,

Table 8. *Estimated Agricultural Production, Acreage, and Value*[1] (1954)

Crop	Acreage (in 1000 acres)		Production (in 1000 tons)		Value (in 1000 £)[2]	
	African	Non-African	African	Non-African	African	Non-African
maize, millet, sorghum	3,329	85	903	32	15,474	648
root crops[3]	1,266	·	2,542	·	23,387	·
beans and pulses	716	15	231	5	4,133	146
bananas	419	·	507	·	4,981	·
rice (paddy)	167	1	58	1	1,332	13
wheat	41	15	11	9	260	248
groundnuts	193	3	25	1	1,453	25
coffee (hulled)	93	29	13	9	5,120	4,607
tea	·	13	·	1.5	·	843
sugar[4]	3	8	·	12	·	89
non-indigenous fruits & veg.	9	3	7	2	99	29
sesame[5]	55	·	6	·	284	·
copra	78	53	7	4	380	196
sisal	·	451	5	167	276	12,621
seed cotton	287	1	55	1	3,495	6
tobacco	7	7	2	2	416	384
papain	·	3	·	·	·	37
pyrethrum	·	3	·	0.5	·	115
Total	6,663	690	4,372	247	61,090	20,007

[1] The figures in this table are rough estimates, and may be quite inaccurate. They were made 7 years ago. See also: IBRD, Table 42, p. 227. See Bibliography.

[2] Including consumption by producers.

[3] Total production calculated on the basis of dried roots.

[4] Excluding production of sugar cane by African farmers for direct consumption.

[5] Excluding acreage for hedge sisal.

Source: UN-Ninth Report of the Committee on Rural Economic Development of the Trust Territories. T/1544, June 1960.

tea, and sisal were the prerogative of the large estates. Since then pyrethrum production on African farms has overtaken that of the estates. In 1960 African production of hedge sisal accounted for 6 per cent of the total.

The first attempts at growing tea were made on small African farms in the Usumbaras and the Southern Highlands. In 1962, peasant tea amounted to 178 acres, and the aim is 2,469 acres by 1969.

The fact that African production now dominates the agricultural picture, accounting for an increasing part of market supply and exports, and spreading to crops which hitherto were exclusive to estates, does not imply that the quality of land use on African farms is improving. One of Tanganyika's most difficult problems is that expanded production for the market has had very little effect on methods of cultivation. The cultural development necessary for this has not yet gone beyond the early stages. Additional land is cultivated by age-old methods, and more, but inferior, cattle are kept. This leads on the one hand — because of the large number of African farmers — to considerable production increases, and on the other to the destruction of the natural vegetation, the reduction of the remaining forests, over-grazing and, finally, to erosion. Reserves of fertile land which should be preserved for future generations, are being destroyed with little benefit to the people now, who could make a much better living by a more judicious use of the areas already cultivated. Increased yields, the most important step towards a consistently developing agriculture, are attained but rarely. A comparison between the current situation and the situation as reflected in publications from the period of German colonial times gives the impression — admittedly difficult to substantiate — that very little has changed in methods of cultivation, plant-husbandry use of fertilizers, erosion control, etc. Yields per acreage have apparently remained the same, in spite of the example of the estates and the advice and help practically forced upon the African farmers by agricultural officers. Where higher yields can be confirmed, they are usually the result of extraneous influences, such as better cotton seed or, in the case of manioc, higher quality plants.

In Tanganyika the hoe is still the essential tool of farming. Extensive areas — if not the major part of cultivated land — are farmed in an intermittent way known as *shifting cultivation.* A piece of land — usually just large enough to cover the farmer's needs and to raise some produce for sale — is reclaimed from the bush. Maize, sorghum, millet, manioc, groundnuts, beans, etc. are usually raised in a mixed cropping system (see Table 9). After a few years of farming, the field is abandoned to grass and bush. In some places cultivation is limited to small fertile plots of land, i.e. termite mounds. Normally the soil is cultivated in cycles: it is farmed for 2—8 years, and then left fallow for 6—20 years.

Shifting cultivation is an appropriate system of farming in remote, thinly populated areas and one which adapts itself to the natural conditions and economic level of tribal subsistence production. It can therefore be termed "balanced exploitation". A system of shifting cultivation is

Table 9. *Some Results of the Sample Survey of African Agriculture, Tanganyika,* 1950 [1, 2]

	Central Region		Eastern Region		Southern Region			Southern Highlands		Lake Region
	Dodoma Mpwapwa	Kondoa	Mahenge	Kilosa	Lindi	Massia	Kllwa	Mbeya	Iringa	Sukumaland
Persons/household	4.02	4.54	4.88	4.17	3.81	4.08	3.86	4.81	5.08	4.20
Crop acreages per household										
Total	4.96	5.51	1.40	4.57	2.43	1.77	1.17	2.22	3.12	5.71
Thereof ... acres are cropped with the following crops	4.92	4.62	1.09	3.99	1.39	1.02	0.95	1.83	2.91	4.03
Production in 100 lbs per household										
maize	4.6	9.2	.	41.7	.	2.9	2.5	14.0	29.7	.
sorghum	14.7	26.6	1.3	20.9	14.6	9.9	9.4	0.9	1.5	23.1
millet	21.9	32.0	.	.	.	.	.	6.5	1.0	4.1
groundnuts	1.0	.	.	.	.	.	.	.	.	.
other pulses	0.3	1.9	.	0.2	1.6	.	2.1	1.0	0.5	.
seedcotton	.	.	1.5	4.0	.	.	.	.	.	1.6
rice	.	.	20.9	8.9	11.0	4.3	13.6	0.9	3.3	1.8
Stock per household										
Cattle	6.8	3.5	0.1	0.1	.	.	.	5.0	4.4	5.6
Goats	2.9	4.8	0.1	2.1	0.1	0.3	.	0.8	1.2	3.1
Sheep	1.4	1.8	0.2	0.2	.	.	.	0.5	0.7	3.4

[1] For information on survey methods, representation and significance see source. Significance of data is low.

[2] These figures are for the harvest of one season only. The survey was done 12 years ago. Since that time production of cotton in Sukumaland and of cashew nuts in the Southern Region has increased significantly.

Source: Report on the Analysis of the Sample Survey of African Agriculture, 1950. Revised. E. A. Stat. Dept. 1953.

not, however, conducive to achieving the bigger yields necessary in the face of an increasing population and a growing market. It is hardly possible to obtain higher yields per acre. Shortening the length of time in which the soil is kept fallow by prolonging the season of cultivation leads to loss of soil fertility or even to erosion. "Balanced exploitation" thus becomes "soil mining". The African farmer who has traditionally practised shifting cultivation has little understanding of soil fertility. The use of fertilizers and the cultivation of fodder crops, are almost unknown to him. When a given plot of land is no longer fertile the farmer moves to new ground, so that finally large areas of noticeably exhausted land are the result. The use of fire plays a disastrous role in this. It is indispensable in the continuous clearing of the bush, but unfortunately it is not always kept under control. Large annual bush fires result in the destruction of tree vegetation, scrub, grass and other plants, and in the lowering of the humus content of the soil.

A more developed form of land-use in Tanganyika is *semi-permanent* cultivation, in which crops are grown in succession, after which the land is left fallow for a period. The distinction between shifting cultivation and semi-permanent farming is that the fallow period of the latter is not long enough to allow the bush to recover. The naturally regenerating grass usually serves as pasture. In some places, such as Sukumaland, semi-permanent cultivation goes with careful soil husbandry. Ridging is customary, resulting in thorough mixing of the soil and excellent protection against water erosion. In spite of this, the system does not by any means ensure the preservation of soil fertility. In general, the same crop or the same mixture of crops is grown on one field until decreasing yields force the farmer to leave the field fallow. There is plenty of manure in the "bomas", but it is not spread on the fields. In this respect Tanganyika contrasts unfavourably with neighbouring countries. Manure is a commodity much sought after in Kenya. From Rhodesia it is reported that approximately one-third of the farmers use manure. In those areas of Tanganyika where semi-permanent cultivation is practised, manure is used in so far as new fields are prepared, preferably on locations of old settlements or "cattle-bomas".

Permanent cultivation is also practised in some areas: along Lake Victoria, in the Usumbara Mts., and on the slopes of Mt. Meru and Mt. Kilimanjaro. It is often practised with the same care given to horticulture. Nevertheless, the yields are low. Mixed farming, i.e. manuring of crops and fodder crop cultivation, is limited to Ukara Island on Lake Victoria, a few valleys in some mountainous areas and the banana-coffee farms of the Wachagga on the slopes of Mt. Kilimanjaro.

Even though the way the soil is cultivated under the systems of semi-permanent and permanent cultivation cannot be much objected to, productivity remains extremely low. Usually, from the standpoint of industrialized countries, the amount of effort put in bears no relation to the returns. According to N.V. Rounce, tilling one acre in the Sukumaland ridge type of cultivation requires approximately 168 hrs., plus 32 hrs. for the ties; added to this is the work necessary for planting, weeding, and harvesting. All in all, one acre of land requires at least 400 hrs. of manual work. Against this, the gross yield per acre is between 100 and, at the most, 300 shs. Tilling with a hoe is a lengthy process. The best times for planting are not kept, partly from negligence or tradition, partly because hoe-tilling is such an effort. In general the farmers first plant their subsistence crops. Cash crops are planted afterwards, and usually too late [1]. For this reason, the harvest is hardly half of what it could be if planted at the most favourable time.

To go over to permanent land use with higher yields per acre, which is bound up with the use of animals or tractors when cultivating the land, is admittedly difficult. A cardinal point would be the development of crop rotation, together with the planting of fodder crops and the use of manure. All of this presupposes far-reaching cultural changes. Rational fodder production is useless without a good stock of cattle and necessitates long-term planning, constant care in feeding and looking after the animals, a knowledge of animal husbandry and, most important, a willingness to give up certain customs and traditions. Some of the more intelligent farmers are prevented from trying to reach a higher level of land use, because they are too dependent on their neighbours, lack the know-how and examples to follow, or because they do not wish to risk any possible savings or to make the extra effort when the results are not at all certain. Furthermore, we do not yet know enough about the best way to apply crop rotation. In general, research done in this field has not yet come up with satisfactory answers.

However, this does not in any way mean that improvements in African land use could come about only by crossing a certain "technological threshhold" which is beyond the capability of individual farmers. Numerous improvements, the usefulness and profitability of which have already been demonstrated, are possible almost anywhere. Yet they are adopted only gradually, if at all. This is especially true of measures to check erosion. Where terracing or ridging are not already customary, they are rejected. In exceptional cases only do the sowing time, the seeds and plants, the intervals between rows, weed-control, the tending of the crops etc. conform to what might be expected and to what the farmer could manage without too much

[1] In the cotton region of the Wasukuma this is true with certain reservations. At the beginning of the rainy season some maize is planted. Otherwise the cotton fields are tilled first. Increasing numbers of farmers are concentrating on cotton and purchasing foodstuffs.

additional effort. According to DE SCHLIPPE, the working day in the Congo averages 5—6 hrs. during the season[1]. In neighbouring Tanganyika the situation is probably similar. There are some "peak working periods" during which the available labour is fully occupied. With the exception of these few weeks, there is ample time to work on improvements.

The situation on African farms varies from place to place. Local differences arise due to climate, soil and market conditions. Tribal traditions account for widely divergent cultivation practices, even when the natural conditions are identical. Some tribes are noted for their rapidly growing interest in commercial production. In others, efforts to introduce innovations have as yet met with no success. Really productive African agriculture is concentrated on a few types of farms. In order to give a clearer understanding of the situation, three of these types of farms will be described below in greater detail. The case of Ukera Island shows traditional farming at its best. In addition, there are some notes on the use of fertilizers, to round off the picture.

3. Some Specific Information about African Farming

a) Cotton Farms in Sukumaland

Practically all of Tanganyika's cotton production, worth about £ 7 million, comes from African farms in the Lake Region, with Sukumaland at the centre. The number of farmers exceeds 250,000. Numerically, and from an economic point of view this is undoubtedly the most important commercial type of farm in Tanganyika.

Several studies have been made of the agricultural situation in Sukumaland (MALCOLM, ROUNCE, WRIGHT, PEAT, BROWN). COLLINSON'S study contains more recent data on the way the farms are organized in Bukumbi and Usmao, two districts which might be considered typical of Sukumaland. The figures in Table 10 refer in each case to farms with average yields.

The Sukuma farm has two separate branches of farming: Crop cultivation which is done by the family and is the main source of income, especially of cash; and animal husbandry, the yardstick for wealth and social standing, which provides some security in times of want and old age, and is the family's source of animal proteins. 6—7 acres are planted annually, i.e. approximately half the acreage available to a family. The rest is kept fallow. Cotton is grown on 3—4 acres, subsistence crops on the remainder. These usually consist of a variety of crops, such as maize, manioc and beans. Rice and sweet potatoes are grown separately. The various crops are not grown in rotation, but rather every year on the soil most suitable for each. When the yields fall off, manioc is planted. This grasses over and leads to the field being kept fallow for some years. Apart from the crops mentioned, hedge sisal is of importance. It is processed by hand.

Work on the fields starts some weeks after the rainy season has set in (October/November). In central Sukumaland it is done exclusively by the hoe and in ridges.

---

[1] DE SCHLIPPE, P.: Méthodes de Recherches Quantitatives dans l'Economie Rural Coutumière de l'Afrique Centrale, Bruxelles 1957, P. 62.

Table 10. *Organization of Cotton Farms in Sukumaland*

	Sample 1961 Bukumbi	Sample 1962 Usmao
Number of Farms[1]	13	39
Size of Family	9	8
Livestock Units[2]	7.4	8.2
Land Use		
Total arable area, acres	12.23	14.10
Acreage cropped, acres	7.19	6.11
thereof cotton, acres	3.73	3.06
paddy, acres	3.46 (paddy and other crops)	0.61
other crops, acres		2.44
Yields		
Cotton, lbs/acre	347	586[3]
Cotton, shs/acre	189	296
Other Crops, shs/acre[4]	165	152
Economic Balance Sheet, shs/farm		
Sales cotton	703	947
other crops	64	603 (other crops and home consumption)
Home consumption of crops	314	
Livestock products, sales	165	495 (sales and home consumption)
Livestock, home consumption	503	
Total value of agric. prod.	2005	2045
Expenses[5]	345	326
Accounting profit	1660	1719
Productivity		
Labour available in man equiv.[6]	3.11	3.18
Labour use in crop cultivation	1.08	1.08
Labour use as p.c. of labour available	33%	34%
Accounting profit per man equiv. in shs	534	540

[1] Only farms of the average profit group.

[2] Key to livestock units: cattle over two years: 1.0, cattle 1—2 years: 0.75, cattle under 1 year: 0.25, goats and sheep: 0.20.

[3] Of total sample, not only average profit group.

[4] Without paddy.

[5] Mainly cattle purchases.

[6] Men, 19—60 years: 1 M.E., women, 19—60 and men 15—19 and over 60: 0.67 M.E., women 15—19 and over 60: 0.5 M.E., children 10—15: 0.25 M.E.

Source: Collinson, M. P.: Report on Bukumbi Survey. Mimeo. 1962. — ditto: Farm Management Survey No. 2. Usmao, Sukumaland, 1963. Mimeo.

Oxen and plough are being used more and more in the border areas of the north and south. Very occasionally one may find a tractor[1]. The planting of cotton takes

[1] In 1963 they numbered 200 or more. The spread of the plough is accompanied by an increase in acreage, a deterioration in the quality of cultivation, more erosion and lower yields per acre. The main advantage is that the work is made easier.

place in December/January. The harvest is brought in during the dry season, lasting from the beginning of June until October. Average annual rainfall amounts to 30 inches, 90 per cent of which falls between November and May. The yield of cotton per acre in the two inquiries is quoted at about 347 lbs and 586 lbs. A yield increase of between 50 and 100 per cent would be possible if planting and weeding are carried out at the right time. Higher yields per acre through better cultivation are not as yet noticeable.

A medium-sized farm keeps 7 or 8 head of livestock. In Bukumbi the average is 11 head of cattle, 4 sheep and 4 goats. One can assume the actual number of livestock to be greater. Some animals are kept in more distant grazing areas. Crop farming and animal husbandry are not integrated. The feeding of the animals and the use of manure are rare exceptions.

The animals live on grazing, fallow land, and crop refuse. About 2.5 acres of grazing are available per head of livestock. There is no check on grazing density or on protection in areas liable to erosion, etc. As against the individual use to which cropped fields are put, there is the general and uncontrolled use of grazing and fallow land. The grazing is overgrazed and very much less productive than it could be if more judiciously used. There is not enough fodder to last through the dry season. The practice of keeping stocks of hay as fodder reserves is not known (nor is the sickle). The livestock ist raised on a rhythm of gaining weight during the rainy season and losing weight during the dry season.

The animals are kept primarily for the purpose of supplying meat for the farming family. Some milk and meat is sold. The income from cotton has raised demand among farmers with little livestock. For Usmao, Collinson estimated that, in spite of the poor quality of the livestock, cattle show an extraordinary high return per head. He assesses the average herds at about £ 90 and the net returns at about £ 22.9 — excluding the costs of herding — which corresponds to 25 per cent rates of interest. However, 1962 was a particularly good year for grazing.

Collinson estimated the average gross yield at about 2000 shs, with a wide margin between more and less successful farms. About one-half of the gross production is sold. Expenses are small and consist mainly of the purchase of additional cattle. The Sukumaland farmers stock up their herds with cattle from the poorer Central Region. After deducting expenses there remains a farm income of about 1600 shs, i.e. about 500 shs per male worker and year. The hire of labour is widespread. In Usmao 77 per cent of the farmers employ occasional labour. Altogether, these perform 11 per cent of the total work.

Table 11. *Productivity of Hoe Cultivation in Sukumaland*
Usmao 1962. Average of 90 farms.

	Cotton	Rice (paddy)[2]	Maize
Yield in lbs/acre[1]	586	1540	422
Yield in shs/acre	296	427	124
Labour use in days per acre[3]	55	127	35
Return per labour day in shs	5.38	3.36	3.54

[1] Data concern the favourable year 1962.

[2] Irrigated rice and hill rice. According to Collinson irrigated rice would give more favourable results.

[3] Man days of 8 hours.

Source: Collinson ibid.

Crop cultivation employs only one third of the available labour capacity. Even during the three peak months of farm work the labour employed is utilized for only 62 per cent of the potential working time. Table 11 gives an idea of the incomes derived from the three main crops of Sukumaland in hoe-farming in 1962, which was a relatively good year. Rice shows the greatest gross returns of 427 shs per acre, as against cotton with 296 shs and maize with 124 shs per acre. However, cotton shows better returns per working day, namely about 5.38 shs, as against 3.36 shs in the case of rice and 3.54 shs with maize.

b) Coffee-Banana Farms of the Wachagga

Coffee, bananas and maize are the chief crops grown on the fertile, well watered slopes of Mt. Kilimanjaro. The relative importance of each crop depends on the height at which it is grown and, therefore, on the amount af rainfall. Below the forest belt it is mainly beans, maize and bananas that are grown; following this there is a broad belt, of particular importance commercially, of coffee and banana farming, with bananas predominating in densely populated areas. Maize is grown on the lower slopes, where there is less rain.

In the course of five decades an important and particularly interesting type of farm has developed in the coffee-banana belt. These are small farms of between 1 and 4 acres, on which almost always mixed coffee and banana crops are grown (see Table 12). In addition, a few miles lower down the mountain these farms grow a plot of maize. Most of the farms keep some cattle and goats stabled throughout the year. With the increase in population the previously communal grassland has been broken up to be replaced by coffee and banana.

The animals are fed with the leaves and stalks of bananas. Grass and hay is brought from the lower steppes, the women carrying bundles of it for miles on their heads. The goats feed on the leaves of hedges surrounding each holding. The main purpose of this type of animal husbandry is to obtain manure.

The various farm activities are integrated to a remarkable extent. Bananas supply food, the raw material for the brewing of beer, some cash, fodder for cattle, and mulching material. The manure for both coffee and bananas comes from the cattle, which also supply animal proteins.

Planting is arranged according to the sunlight requirements of each crop. Light permitting, beans are grown nearest the ground; above them come the coffee trees, which in turn grow in the shade of bananas. The smaller the farm the more preference is given to bananas, because under present methods of cultivation they yield in terms of money the biggest gross returns per acre.

Beck's study in 1961, a particularly bad year, shows that an average 3-acre holding of this type yielded gross returns of 1880 shs. The main outlay was the wages bill for outside labour. One third of the families derived an income from outside the farms, usually by working on the estates. The average family income in 1961 was quoted at 1463 shs. According to another more widespread but less systematic enquiry in 1959, the returns from coffee and bananas were considerably higher even on smaller farms.

As everywhere in Tanganyika, correct methods of farming on the coffee-banana farms of the Wachagga could lead to a cheap and strong rise in production. The low yield of 0.5—0.8 lbs per coffee tree is not at all a necessary consequence of the presence of bananas, but is mainly due to negligence. But this type of farm is distinguished by the fact that it shows numerous individual attempts at farm improvements. In contrast to all other forms of farming in Tanganyika, one can note an unmistakable increase in the yield per acre. On many of these farms husbandry of coffee trees is equal to, and even better than on the estates. Use of

Table 12. *Organization of Coffee Banana-Farms at Mt. Kilimanjaro* [1]

	Machame Central Area 1961	Agricultural Dep. 1959
Number of farms	100	·
Size of family	7.5	·
Livestock units per farm	1.2	·
Land use per farm		
Total area cultivated, acres	3.2	1.27
Thereof coffee	2.7	·
banana	2.5	·
beans	1.2	·
maize	1.8	·
other crops	0.2	·
Mixed cropping of coffee and banana occurs on ... acres	2.2	
Coffee trees per farm	845	588
Yields		
Coffee, lbs/tree	0.6[2]	0.85
Coffee, lbs/farm	430	500
Banana bunches per farm	309	970
thereof sold	124	·
Economic balance sheet, shs/farm		
Receipts		·
coffee	723	850
banana	337	2650[5] (banana, other crops, home consumption)
other crops	120[3 4]	
Home consumption	700	
Total value of agric. prod.	1,880	3,500
Expenses (without wages)	457	·
Farm income	1,323	·
Wages for non-family labour	325	·
Non-agricultural income[6]	465	·
Family income[7]	1,463	·
Productivity		
Labour available, M.E.	2.1	·
thereof non-family labour	0.5	·
Family income per man unit of family labour, shs	914	·

[1] Data about Machame are an arithmetic mean of a systematic sample of 100 farms. Data of the Agr. Dep. were obtained in an unsystematic wide spread survey. They are less reliable.

[2] ⌀ 1960 and 1961.

[3] Incl. change in capital invested.

[4] 1961 harvest was exceptionally poor.

insecticides is customary. Interest is growing in copper spraying and mineral fertilizers. Most of the farmers buy some mulch (steppe-hay) in addition to their cattle manure.

Worthy of note, too, are the beginnings of a rational milk production. The income from coffee has resulted in a greater demand for milk. The price per gallon is 3—4 shs. Some farmers are buying Jersey cows from the estates in order to sell milk to their neighbours. The annual milk yield per exotic cow lies between 400—600 gallons. Some farmers buy additional bran and pyrethrum cake; they also grow grass for fodder, either in terraces between the coffee or in separate plots. If the cows are properly looked after, fodder grass shows greater monetary returns per acre than either coffee or bananas. Hence grass becomes another cash crop. The farmers are encouraged to grow their own grass by the fact that the purchase of hay from the steppes carries the danger of introducing East Coast Fever. Disease-prone exotic cows are kept healthy by using home-grown fodder and stable-feeding all the year round.

The fact that Wachagga farming is commercialized to such an extent has also led to changes in their attitude to land tenure. Land is bought and sold. It serves as security against government loans. It can be considered the private property of individual farmers.

c) Wheat Farms in the Northern Region

During the past decades a number of wheat growing African farms have developed in the districts of Arusha and Mbulu. In the Northern Region there are 270 farmers with more than 50 acres, and 91 with more than 100 acres of arable land. The soils are volcanic, with an annual rainfall of about 30 inches. A typical larger farm of this kind consists of about 200 acres, with 100 acres of wheat, 40 of beans and 60 of maize. The yields per acre vary from year to year, but average about 6 bags of wheat, 3.5 of beans and 7 bags of maize. At 50 shs/bag for wheat, 75 shs for seed beans and 30 shs for maize, the gross returns of a 200 acre farm amount to approcimately 53,100 shs.

The fields are tilled in April/May. The wheat is broadcast, the maize planted by hand. The beans are sown with the aid of special sowing machines. In some cases European estates do some of the work under contract. In the late fall, wheat is sown a second time on part of the land in the hope that the "short rains" will be sufficient for a harvest.

The farms are equipped with 1—2 tractors, a few have a combine, a maize-picker and a truck. Contract work on other farms is an important source of additional income. The larger wheat farmers often plough the fields of small farmers in the vicinity. The price for ploughing an acre is 40 shs. Combining costs 6—8 shs per bag or 40 shs per acre. Permanent farmhands on the wheat farms are limited to 2—4, and these are often members of the family. When more labour is needed, i.e. for tilling, hoeing of the beans, harvesting etc., temporary labour is taken on from among peasant farms in the neighbourhood.

Some of the larger farmers are former chiefs who invested their compensation money in tractors. They first engaged in contract work, but later switched to farming on their own. As the native land law does not permit purchase of land,

---

[5] Only banana, 2.73 shs/bunch.

[6] In both surveys, 30 p. c. of all families receive non-agricultural income.

[7] Without rent for house. Beck calculates rent per family and year as 420 shs.

Sources: a) Beck, R. S.: An economic study of coffee-banana-farms in the Central Machame area. 1961. Mimeo. 1963.

b) Agr. Dep. District Book, Moshi.

some of the big wheat farmers are in the habit of leasing land from smaller farmers for one year against the payment of one bag per acre, or every 9th bag. Consequently the rich farmer is the tenant and the poor one the landlord. Lately there has been a trend towards buying up "alienated" land from the Europeans or Asians, in order to acquire rights of tenure over cultivated land.

d) Mixed Farming on Ukara

An exceptional and noteworthy case is the organization of land use on Ukara [1—4]. This is an island in Lake Victoria, situated some miles north of Ukerewe. Its total area is 29 square miles and, according to ROUNCE & THORNTON it receives 56 inches of rainfall, according to PATERSON 60—70 inches. Its population in 1962 was about 18,000. Before World War One, according to the "Deutsches Kolonial Lexikon" (Vol. III. 1920, p. 569), the population was 23—28,000.

Less than half of the area is cultivated, or is suitable for cultivation. There are 620 people to each square mile of the total area, and about 1,300 to each square mile of arable land. The average farm area per family of 4.5 persons is 2—2.5 acres, i.e. approximately one third the area of land farmed by a household in neighbouring Sukumaland. Moreover, the soil is mostly sandy, of low fertility, and overworked due to continuous cultivation. As a results of the isolated position and tribal and family ties, there is little migration to neighbouring mainland areas where the soil is quite fertile and the population small. Thus there is a shortage of land on Ukara unknown anywhere else in the Lake Region. This shortage of land has led the Wakara to practice crop rotation and manuring, to plant fodder crops, and to farm on individual holdings where trees and grassland are husbanded. These features cannot be found to this extent in any other farming areas in Tanganyika [5].

Crop Rotation. According to a census of 1945 the Mkara plants an average of 1.19 acres of millet, 0.67 acres of manioc, 0.28 acres of rice, 0.16 acres of sweet potatoes and small amounts of sorghum, finger millet and vegetables. In addition bambara nuts are planted between these crops, green manuring with a legume is practiced (crotalaria striata) and, on the Lake shores, fodder grass is cultivated. Rice, sweet potatoes, and fodder grass are cultivated more or less as permanent crops on especially moist soil. For the other crops ROUNCE & THORNTON describe a crop rotation of the 1930's which existed in 1945 and 1956, and in 1963 was still the standard method [6].

1st Year

August	Application of manure,
September/October	Millet is planted, crotalaria striata is sown after the first hoeing, i.e. when the millet is 12 inches high,
November/December	Growth of the millet and crotalaria striata,
January/February	Millet is harvested,
March/June	Growth of crotalaria striata to a height of 4—5 feet.

[1] This chapter is mainly based on information provided by ROUNCE & THORNTON, LUNAN & BREWIN, MALCOLM, PATERSON, RATSEY (see bibliography).

[2] See also: Agricultural District Book, Regional Agricultural Office, Mwanza.

[3] East African Medical Survey, Annual Reports 1951, 1952, & 1953.

[4] Two visits by the author in 1963.

[5] In the other example of mixed farming, the coffee-banana farms of the Wachagga, the leading crop is a permanent one.

[6] According to ROUNCE & THORNTON the harvesting is done in January/February. In 1963 the harvesting of millet began in mid-April.

2nd Year

July	Crotalaria striata is used as green manure during the dry season,
September/October	At the beginning of the rainy season the millet which has fallen germinates. A second crop of millet begins to grow.
November/January	Bambara nuts or groundnuts are planted between the rows of growing millet,
February	Millet is harvested,
March/May	Growth of the groundnuts,
June	Groundnuts are harvested.

3rd Year

July/September	Application of manure,
September/October	Self-sown millet germinates,
November/January	Growth of the millet,
January/February	Millet is harvested,
February/April	Sorghum is sown or planted,
June	Sorghum is harvested.

Manioc, a crop introduced in 1940, is planted between the millet, sometimes mixed with groundnuts, as a tumble-down fallow. Fallow land is seen only rarely, but sometimes the older, grassed-over manioc fields are referred to as fallows. In valley areas, since 1920, more and more rice is grown outside the above described rotation and is irrigated with brook water. The rice is sown in seed-beds and then transplanted. Rice is almost always grown for sale. The other crops are used by the households themselves.

In 1945, the average Mkara household kept 2—3 head of cattle and 1—2 goats or sheep. The need for manure and fodder forced the Wakara to integrate crop cultivation and animal husbandry. Little grazing is available. According to Rounce & Thornton there are only 0.9 acres of grassland for each livestock unit, the grass of which is partly used as roofing material.

Livestock Feeding. The following products are used as fodder:

1. Crop refuse, especially the straw from millet, rice and sorghum.
2. Groundnut leaves, sweet potato and manioc leaves.
3. Picked and dried sorghum leaves.
4. All weeds. The women and children carry baskets with them while cutting the weeds. The weeds are carried to the cattle shed.
5. The leaves of 11 different trees. The trees are owned privately. Leaves and twigs are regularly harvested and used for fodder, bedding, or in the construction of huts. The use of trees is commercialized. Lunan & Brewin report the case of a tree owner demanding 10 shs for one cutting of leaves and twigs.
6. On the lake shore fodder grass is planted and cut several times a year. It is irrigated by trenches and ditches bringing water from Lake Victoria.
7. Extensive areas of little fertility in the center of the island are grassland, grown over with Trichopterys uagerensis. This relatively poor grass is used as a roof covering. After the harvest it is grazed. Another grazing period follows immediately after the first rainfall as long as the grass is not higher than 12 inches.
8. Occasionally rice is planted in June, i.e. at the beginning of the dry season, and is used as fodder during the period in which little other fodder is available.

Animals are slaughtered, especially at the beginning of the dry season, in order to correlate the demand and supply of fodder. Male calves are not raised.

The supply of fodder is so small that animal husbandry must be limited to its major purpose: to provide manure and essential proteins (milk). Milk production, however, is very low. The shortage of fodder has led to the development of dwarf zebus. According to RATSEY, an average cow weighs 4 cwt, bears a calf only once in 2—2½ years and supplies about 1½ lbs of milk daily during the seven-month lactation period. Since the Wakara consume fish, cattle husbandry is probably less important as a source of protein than of manure. Thus, the major function of cattle husbandry is not to provide capital for emergencies or for old-age as in neighbouring Sukumaland, but to preserve the fertility of the soil. The shortage of fodder prevents the building up of wealth in livestock.

The Use of Fertilizers.

1. Cattle are kept in sheds practically the whole year round. They take up half of the huts and are put outside only to drink and occasionally to graze. Sometimes they are kept tethered. In the huts the cattle are kept in 4-ft. deep stalls, the sides of which are banked by stones. Thus, no liquid manure is lost. Feeding and bedding are done twice daily. As soon as the stall is full, the manure is removed and stored in piles in front of the huts. Three times a year, the manure is brought to the fields in baskets which are carried on the head. ROUNCE & THORNTON estimate the supply of manure at 4 tons/acre annually. They calculate that the millet yields are 30—40 per cent higher than on unfertilized plots.
2. The bedding serves to return to the soil not only nutritive elements extracted with the crop. By using leaves and twigs of all the trees and bushes of the island, as well as by using for bedding the dried grass which has served as roofing material, nutritive materials found outside of the cultivated fields are concentrated in the manure and returned to the land.
3. The plant which is used as a green manure is a deep-rooting legume containing nitrogen-building bacteria. The green crop, which may amount to 9 tons/acre, is dug under during the dry season. If there is a massive growth it may be partly cut and carried to another field.
4. Leaves and twigs, in addition to being used as fodder and for bedding, are transported to the fields and applied as green manure.
5. Nurseries for rice are fertilized with household ashes.
6. Urine is added to the manure heaps. According to unconfirmed information obtained in the summer of 1963, the farmers prefer to relieve nature on their own fields.
7. Fertile undersoils which are dug up when digging trenches for the irrigation of fodder grass along the lake shore are used to fertilize the top sandy soils.

Erosion Control. In addition, the Wakara have developed their own methods of erosion control:

1. The legume used as green manure grows very slowly and ripens only after 9—10 months. Thus it covers the soil for long periods and protects it from erosion.
2. In some instances tie-ridging is practised. In general before tilling, square depressions are dug out, varying in size between 1 and 9 sq. yds. Manure and leaves are spread in the squares. Soil from the unfertilized center of the squares is dug out with a hoe. In this way the manure reaches the core of the ridges surrounding the square. The depressions are apparently dug only at the beginning of the rotation cycle. In the course of time they become levelled off so that only a sloping of the surface area is visible.

3. In a few cases stone terraces have been constructed. They serve principally to dam up brooks.
4. Grass is sometimes planted to reinforce the banks of brooks and erosion gullies.
5. Scheven reports that soil which has been washed away is carried back in baskets, and that flat rocks have been covered with earth, so that crops can grow on them.

The measures taken by the inhabitants to control erosion have not been sufficient. In the areas of low fertility where the grass is used for roof coverings, long and deep erosion gullies have developed. The local administration has temporarily financed the construction of terraces. Where these terraces exist erosion has been kept under control.

Attitude concerning Land Ownership. On Ukara, land is the private property of the male head of the family. With the exception of a few poorly-maintained communal grazing areas, this applies also to grassland. Great importance is attached to the ownership of fodder. Cattle are sometimes muzzled, in order to prevent damage to the neighbour's crop. Cattle may not be driven over the neighbour's property without his consent. Upon the death of the owner, the property is usually divided among his sons. Thus there are many small holdings. Few fields are larger than 1/10 acre, some are only 20 sq.yds. The borders of many plots are marked with border stones. Land is bought and sold. In 1959 — according to Scheven — properties the size "of classroom" changed hands for 400—800 shs.

Summary about Ukara. Though the agriculture of Ukara is remarkable, it is by no means a model of agricultural development, but a relic of a more poverty-stricken past. The careful soil husbandry is not enough to raise incomes and living standards. It is a result of an insular pressure situation which existed in Ukara before the European colonization of East Africa [1]. The Wakara clearly work more than other tribes. Yet their poverty is great. They lack a proper cash crop. The only things sold from the island are rice, dried manioc and roofing grass. They may reduce the loss of soil fertility by the use of manures, but they are not able to prevent it. They are proud of the fact that visitors praise their system of land use. Yet this inherited system is part of the traditional order of a once forceful tribal conformity, and is breaking down, as are the tribal ties. There is no indication of any willingness to make the step towards any further development, which is quite possible, for example by using mineral fertilizers. More profitable outlets are seen in:

— cultivation of rice in the valleys,
— fishing for the market in Mwanza,
— seasonal work on the thriving neighbouring island of Ukerewe,
— cultivation of cotton on the mainland, with seasonal migration there.

As the living standard slowly rises — this is apparent in the clothing of the people — so soil husbandry is deteriorating. Terraces fall into decay. Areas of fodder grass which have been ruined by the level of Lake Victoria rising are not being rebuilt to the previous extent. Terracing which was financed for a time by

[1] Cultivation practices on Ukara apparently developed without European influence. The first European visitors to the island report "an amazing quality of cultivation", "well-kept fields", "use of leaves for fodder", "lake shore irrigation", "manuring" and "erosion control via small ponds and pits". (See: Baumann, O.: Durchs Massailand zur Nilquelle, Berlin 1894, p. 48; Schweinitz, Graf v.: Deutsch-Ostafrika, Berlin 1894, p. 168; Meyer, H.: Das deutsche Kolonialreich, Leipzig 1909, p. 291; Schlobach: Die Volksstämme der deutschen Ostküste des Victoria Nyansa, in: Mitt. aus den deutschen Schutzgebieten, Bd. 14, 1901.)

the local administration is not continued. The change-over from the traditional "beehive house" to the round hut has reduced the demand for roofing grass. This has led to private farming extending over previously communal pasture land. This, in turn, reduces the shortage of land. Land has become cheaper. An increase in farming land, coupled with a decrease in livestock, leads to less manure becoming available. Erosion is on the increase.

Those Wakara who settle on the mainland forget their laborious but fertility-preserving practices and revert to the normal "soil mining" practices of the Wasukuma. Their system of land use is not an example which others follow. Those who have stayed behind on the island, where they have to work from sun-up to sun-down in order to make a poor living, are described as foolish, and their manuring practices as "filthy".

Thus, land use in Ukara is an interesting special case which shows that under the present conditions of economic and cultural development in the Lake Region, only an extreme emergency situation can induce the inhabitants to take up pre-technical measures to preserve the fertility of the soil.

e) Some Examples of Traditional Methods of Manuring

(a) **Chiteme System, Southern Tanganyika.** Bushes are transported to small arable plots of land and burned there. The ashes serve as a mineral fertilizer. It is assumed that the relationship between the land thus fertilized and the bush area which supplies the material is 1 : 10.

(b) **Pit System, Mbeya District.** Grass from fallow land is cut and heaped up. Top soil is then placed on top of the heaps so that they finally reach a height of 2 feet and a cross-section of 1 yard; these are then usually planted with beans. The beans are harvested in June/July. Weeds growing on the plot of land are pulled up and thrown onto the compost heaps. When the rainy season begins, i.e. in Nov./Dec., the heaps are pulled down and the decomposed plant material is strewn onto the soil like ordinary manure. Finally maize and millet are planted. The placing of plant refuse in heaps, the process of rotting down and the distribution of the resulting compost is continued for a time. When the soil becomes exhausted the field is again kept fallow.

(c) **Hole Culture in Songea.** Grass is cut and formed in squares of 2 ft. by 6 ft. The surrounding soil is then hoed onto the squares. This results in a row of such squares, alternating with holes in the ground. Maize, millet, wheat, etc. are planted on these grass squares. Weeds, plant refuse etc. are thrown into the holes and provide compost for the following year. On the spot where the hole was located, a new pile is built, using the soil from the surrounding piles, the plant material in the hole serving as an underlying fertilizer. In this manner the soil can be used for about 10 years before fertility is exhausted and the plot abandoned.

(d) **Changing Living Location.** In and around the living quarters the soil is enriched by human and animal waste. By moving his dwelling the farmer obtains a plot of richly fertilized land. Blohm reports that among the Nyamwezi an ambitious man removes his dwelling 2—3 times during his lifetime.

(e) **Cattle-tethering on the Coast.** The owners of coconut palms — often Arabs — tether cattle to the trees at night or build bomas under palm trees and frequently remove them to other places. In some districts the coconut palms are manured every three years in this manner.

(f) **Ridging.** In Sukumaland, preparation of the soil is carried out in the following three stages: 1. Removal of weeds. 2. The dried weeds are placed in the furrows of the previous year. 3. Soil is piled onto the weeds from both sides, providing an underlying green manure in the middle of the ridge.

4. Animal Husbandry — Peasants and Herdsmen

Small as might be the technical progress in crop cultivation, increasing production for the market has aroused interest in trade, created additional wants and brought about a change in the economic attitude of the farmer. These changes can serve as a basis for further improvements. The example of cotton, coffee, and wheat farmers indicates that various tribes are willing to adopt agricultural innovations as long as these meet the following conditions: they must offer clear-cut monetary advantages, be within the realm of possibility, must not involve too much risk, and should be accompanied by such overwhelming attractions as tractors or combines. The prospects for further improvements in agriculture are therefore not bad. This cannot be said for animal husbandry.

The potential in terms of livestock and grazing is high, but productivity per cattle-unit and per acre is extremely low. For the year 1960, livestock numbers are recorded as 8 million head of cattle, 2.8 million sheep, 4.3 million goats, 141,000 pigs and 16,000 donkeys. Livestock is concentrated in those areas which are not infested by tsetse-fly. In the Lake Region there are 2.6 million head of cattle and in the Northern Region, 1.2 million. 98 per cent of the cattle and practically all of the sheep, goats and donkeys are owned by Africans.

Table 9 shows that the average household — in areas where animal husbandry is possible — keeps 4—7 head of cattle and a few sheep and goats. There is a marked differentiation in the ownership of cattle. The Veterinary Department assumes that 25 per cent of the families own 90 per cent of all cattle. Rounce reported in 1950, with reference to Sukumaland, that almost all of the cattle is owned by one half of the families. Collinson's research on Bukumbi indicates, that 75 per cent of the livestock is owned by the more successful third of all households. It is likely that the concentration of ownership is greater than these figures suggest. Husbanding of cattle does not necessarily indicate ownership. The farmer who can afford to purchase cattle may not be in a position to keep all of them on his location, as is frequently the case with the successful cotton, coffee or pyrethrum grower. He usually leases part of his herd to friends and relatives in distant grazing areas.

The organization of animal husbandry has been the same for as long as man can remember. Open bush, neglected grazings, fallow land and refuse from the harvest provide the basic fodder. The animals are put to grazing late in the morning. At night they are kept in the village "boma" which provides protection from wild animals and thieves. Each day the cattle must cover long distances from the village to grazing areas and watering places. Fodder is abundant during the rainy, scarce in the dry, season. Annual burning of grass and bush regularly destroys fodder which might otherwise be used in the dry season. Fire contributes to the deterioration of the natural vegetation. On the other hand, the use of fire for reducing bush and doing away with useless grass is almost unavoidable. Cultivation of fodder crops and animal feed are practically unknown. In the rainy season the animals gain in weight, and in the dry season they lose weight. Accordingly, raising animals takes a longer time. Only about half of the cows calve annually; of the new-born calves only about 50 per cent survive; in some places it is said to be as low as 10 per cent.

Table 13. *Livestock population* in 1,000 head

	1952	1960
cattle	6,470	7,940
sheep	2,510	2,871
goats	3,530	4,287
donkeys . . .	·	141
pigs	12	16

Sources: a) Annual Report of Dept. of Vet. 1960, p. 14.
b) Statistical Abstracts 1961, Table G. 7, p. 66.

The low productivity of animal husbandry is not a necessary consequence of the natural and economic conditions. Tanganyika undoubtedly has favourable conditions for cattle ranching and it could become one of the most important industries in the country, considering the vast areas of land which receive less than 30″ of rainfall annually. Possible grazing density fluctuates between 3—20 acres per livestock unit [1]. On natural grazing a density of 10 acres per livestock unit can be considered suitable, on well managed pasture a fodder growth of 1—5 acres is sufficient. Experience on some ranches indicates that the usual diseases can be kept under control.

The low productivity of animal husbandry is due principally to institutional features. The traditional forms of crop farming leave little scope for differentiation among farmers. The size of the farm is limited by the family's capacity for labour. The practice of employing wage-earners gives rise to a series of unusual problems: supervision of the workers, organization of work processes, provision of ready cash to pay wages, discharging unsuitable workers, etc. In Sukumaland, for example, it is considered an offence against custom to hire labour, except for seasonal work. In short, arable farming is organized on an "egalitarian" basis. The ordinary human striving towards distinction receives little incentive under such a system. Social standing and material possessions are, therefore, reflected in the number of livestock owned. The owner of many head of cattle is rich. Consequently, livestock is not looked upon only as a means of earning a livelihood, but as a "value in itself".

[1] For the key to livestock units compare table 10.

Ownership of livestock serves in addition to protect the family and marriage. A man wishing to marry is obliged to give the father of the bride some livestock. This custom serves to safeguard both partners in the marriage. If the bride is treated badly she can return to her father, who is not required to return the cattle he received as bride-price. Should the bride be barren or behave badly, the husband can send her back to her father and demand the return of his cattle. These social customs necessarily lead to the hoarding of cattle. A man must be in a position to part with a sufficient number of cattle at the time that his sons marry, or when the bridegroom has reason to demand his cattle back. It is true, money could serve the same purpose. Occasionally the equivalent amount of money is paid instead of cattle, but as a rule money is too abstract a symbol of wealth. It is said that "money produces no offspring", in most places a justifiable statement, since the usual way to save money is to bury it. Farmers generally have a habit of saving, and the farmers in Tanganyika are no exception. The most profitable investment the farmer knows is normally in cattle, both in terms of interests and social standing. Investments in cattle are regarded as the best old age insurance. Thus, the tendency towards continuous quantitative expansion of herds is closely bound up with tradition and custom. Furthermore, the greater the herd, the greater the share in communal grazing. Cattle are bred not so much for the purpose of producing meat and milk as of increasing the herd. Past estimates indicate that efforts to increase cattle numbers have been quite successful. In 1913 livestock were numbered at 4.063 million, in 1931 at 5.022 million and in 1961 at 7.9 million head.

The social significance of owning cattle would be less dangerous if individual rights existed to the grazing, or if there were some institution to regulate the grazing density. This, however, is not the case. Everybody has the right to graze all his cattle on communal land. Private pastures can be found in a few densely populated mountain areas, where this is the wish of the local people. There are no village or tribal institutions responsible for organizing the use of grazing, taking measures towards erosion control, or deciding on growing higher quality grasses. As a result, vast areas are disastrously over-grazed. The perennial grasses, which produce high yields per acre, are pushed out by annual grasses. According to an estimate of the Veterinary Department, a comparison of the land area of Tanganyika not infested with the tsetse-fly with the number of livestock, shows a grazing density of about 9 acres per head of livestock. If cultivated land, forest, mountains, etc. are deducted, there remain no more than 2—5 acres per head of livestock. This could only be sufficient if grazing practices were judiciously organized.

A consequence is the appearance of the so-called "livestock cycle". It begins (1) with a relatively small number of animals, sufficient grazing in the dry season, and a high rate of increase of the livestock. (2) More cattle means less fodder, especially in the dry season. The grass growing is, however, still enough to last through the dry season, i.e. the herd continues to increase. At more or less regular intervals (3) an extremely dry year sets in. Fodder and water are no longer in sufficient supply. Some of the animals die. In the following year the cycle of few cattle, abundant fodder and higher rates of increase begins anew. In some parts of Tanganyika, in Masailand for example, this cycle repeats itself at intervals of about 10 years. During the last dry period — 1961 — half of the cattle in southern Masailand and the western Arusha Chiefdom are said to have died.

The harm done by irrational animal husbandry is not confined to losses during the dry season. Cotton, pyrethrum and coffee are relatively good money earners. One can assume the more successful third of the farmers to save quite a bit. These savings are rarely used for productive purposes. A large part of them is put into cattle. Savings are lost for capital formation because there are phenomena such as the "cotton cycle". Cotton (1) means cash earnings. Some of these earnings are saved and (2) invested in cattle. Hence, there is a concentration of cattle in cotton-growing areas where, in consequence of increased cotton cultivation, grazing has been cut down. This (3) in turn leads to the cattle having less fodder, and (4) in particularly bad years to losses caused by lack of fodder. These losses mean that the farmers have to replace cattle and, to do this, need to earn money, which in turn results in even more cotton being grown. Thus the economic benefit of cotton production is largely consumed by cattle.

Chapter C

Achievements of Tanganyika's Agriculture

I. Production

Reviewing Tanganyika's agricultural development during the last ten years, we find that the standard of husbandry is still very low, measured by what is known and what is economically practicable under existing conditions. It must be stressed, however, that production, and therefore productivity, have increased at a remarkable rate. The resulting picture is one of rapid change from a very low starting point.

For the period 1945—1960 one can roughly assume the national product of Tanganyika to have doubled and the output per head of population to have increased by a half. This is due mainly to an expansion in agriculture. Sisal production rose in the decade 1950—1960 by 60 per cent.

Table 14. *Production and Market Sales of Principal Crops* in 1,000 tons

	1950	1951	1952	1953	1954	1955	1956	1957	1958	1959	1960	1961	1962	1963[3]
I. Production[1]														
sisal	122	145	162	169	178	176	186	185	194	209	205	201	219	220
cotton lint	8,9	8,5	13,9	9,0	18,3	21,5	23,6	30,2	30,7	35,4	32,0	33,0	35,2	·
coffee	16,3	15,4	14,8	15,0	19,4	19,0	22,5	20,7	22,5	22,7	26,2	19,3	26,5	27,7
tea	0,85	1,05	1,10	1,27	1,60	2,00	2,40	2,80	2,80	3,60	3,70	4,38	4,24	4,60
sugar	7,8	8,2	10,1	11,7	10,7	10,4	18,0	18,4	21,0	27,7	28,7	28,7	39,0	55,0
pyrethrum	0,29	0,30	0,24	0,29	0,48	0,67	0,74	0,74	0,61	0,77	1,91	1,27	1,89	2,70
tobacco	1,28	1,24	1,06	0,98	1,59	1,51	1,60	1,60	1,70	2,09	1,78	2,61	2,0	·
II. Sales on local markets[2]														
cashew	5,8	6,4	10,8	10,1	13,2	15,0	16,6	23,4	20,6	28,4	56,3	28,6	55,0	60,0
groundnuts	1,2	5,5	10,9	1,2	4,1	8,6	14,7	15,3	9,7	15,9	22,8	·	16,5	·
rice (paddy)	19,9	21,9	15,6	16,6	16,6	32,1	14,4	27,9	29,9	30,4	34,9	22,9	39,5	·
castor seed	2,0	3,8	7,2	6,1	4,7	7,9	12,0	13,9	6,9	6,2	10,9	12,0	10,0	·
maize	62,5	67,1	48,3	22,9	57,5	121,6	60,0	51,7	67,1	70,1	77,9	55,5	71,8	·
sesame	2,1	3,5	1,9	1,5	2,7	4,5	6,9	6,5	7,0	10,2	3,5	7,5	8,8	·
sunflower seed	5,2	5,6	12,3	2,4	4,6	10,3	7,9	12,6	8,8	4,7	7,3	·	·	·
copra	11,0	8,6	14,9	12,1	12,6	10,9	9,7	8,3	10,9	11,6	9,5	·	·	·
pulses	14,2	24,4	20,8	12,3	16,7	13,6	14,3	14,2	12,8	15,3	24,4	·	19,6	·
wheat	8,1	5,8	5,3	4,9	9,8	4,2	4,8	3,9	6,6	7,3	11,7	·	16,9	·
onions	2,7	2,4	2,2	1,8	2,2	2,7	4,7	3,2	2,3	6,5	8,0	·	12,7	·

[1] The figures given in the various Annual Reports are not always consistent. Here the latest figures are given. The figures for coffee are probably too low, since coffee is smuggeled out to Uganda.

[2] Estimated amounts which are sold on the local markets and recorded for tax purposes. Increasing amounts are probably sold outside the markets in order to avoid payment of the market tax. Both the extent and the rate of increase of marketed produce is probably underestimated in these figures.

[3] Provisional estimates.

Sources: a) Annual Reports, Department of Agriculture, Part I, 1950—1960.
b) Budget Survey 1957—58 until 1962—63.
c) Statistical Abstracts 1961.

Production of cotton, tea, sugar, and pyrethrum increased three or fourfold, that of coffee and tobacco doubled. No reliable production figures are available for other crops. Table 14 contains data on market supplies as published annually by the Agricultural Department. They should be used with care. These data are based on the amounts sold on local markets and, therefore, recorded for tax purposes. To an increasing extent trading is done outside markets, in order to evade the market tax. The real figures for the proportion of production which is sold and the rate at which it increases are probably higher.

Although the quantitative information may be inaccurate, one can probably assume that there has been a considerable increase in sales. The supply of cashew nuts increased almost tenfold. An upward trent can be recorded in the case of groundnuts, rice, castor seed, sesame, wheat and onions, though subject to strong fluctuations from year to year. The trend in the production of maize, sunflower seed, and seed beans is not clear. The fact that the registered supply of copra is stagnating is due less to production lagging behind, than to the rapid increase of the statistically unrecorded trade in fresh nuts.

Table 15. *An Estimate of Animal Production* 1960

	Numbers	tons	Value in mill. £
I. Sales			
cattle[1]	278,436	63,392	2,421
sheep/goats[1] . . .	105,666	3,783	0,143
hides and skins . .	1,214,119	13,915	1,087
milk	—	·	0,173
II. Subsistence Economy[2]			
cattle[1]	1,214,119	276,428	15,600
sheep/goats[1] . . .	2,674,471	95,516	2,257
milk	—	181,581	—
Total			21,681

[1] Liveweight per head of cattle averages 510 lbs, sheep and goats 80 lbs.

[2] This is a rough estimate based on the dried skins delivered and on estimated average weights. Imported cattle from Kenya are not included. It is estimated that there are 3,5 million cows, half of which supply 114 lbs of milk each for human consumption. The estimated consumption in the subsistence sector is evaluated in wholesale prices. The results of this estimate are much higher than the £ 18,4 million which are used to calculate the gross domestic product.

Source: The Role of Livestock in the Subsistence and Cash Economy of Tanganyika. Mimeo. Vet. Dept. 1961.

Tables 15 and 16 contain estimates about animal production and the market supply of cattle. According to these undoubtedly very rough estimates, production of meat for sale and for home consumption affects

1.5 million head of cattle with a total live weight of 340,000 tons and 2.8 million sheep and goats weighing a total of 100,000 tons. The value of this meat production — consumption by producers valued at local prices — is £ 18 million. The total value of animal production, including sales of dried hides, is therefore almost £ 22 million. The above-mentioned study of the Veterinary Department estimates the annual per capita consumption of meat, including offals which are normally consumed, at 50 lbs, and that of milk at 5 gallons.

Table 16 gives some details about the supply of cattle. Sales have decreased rather than increased. The upward trend in the supply of immatures stopped in 1956. In notable contrast to crop cultivation, which is apparently receptive to commercialization and innovations, the supply

Table 16. *Development of Cattle Sales*

Year	Cattle[1,3]		Immatures[2]	
	in 1000 head	price in shs per head	in 1000 head	price in shs per head
1950	212	82	5	46
1951	181	144	11	101
1952	256	137	26	82
1953	282	124	29	63
1954	271	125	45	59
1955	260	133	56	69
1956	261	132	82	64
1957	232	166	61	72
1958	225	169	51	82
1959	245	187	36	93
1960	234	186	34	101
1961	234	158	42	86
1962	217	165	·	·

[1] Local markets; according to the Market reports.
[2] Sales of Persians and lean cattle, usually for further fattening.
[3] Average liveweight of cattle is 510 lbs.

Source: Compiled from the Annual Reports of The Veterinary Department.

of cattle remains stagnant, though prices have more than doubled in the decade 1950—1960. The explanation for this phenomenon is probably as follows: Higher earnings from cash crops have resulted in the farmers consuming more meat themselves, or else trade outside the official markets has increased between farmers and herdsmen.

II. Agricultural Exports and Imports

Because of the low purchasing power of the internal market, increases in production and in sales must be reflected in greater exports. In the decade 1950—1960 the export of cash crops has risen from an average of

260,000 tons in the years 1950—1953 to more than 400,000 tons. Corresponding figures for the value of exports are £ 31 million (1950—1953) and £ 33 million (1958—1960) with a peak of £ 41 million in 1960. In spite of a significant fall in prices for the three most important crops, sisal, cotton, and coffee, the value of exports has increased (Table 17). Note must also be taken of the upward trend in the export of meat products. In 1951 they amounted to £ 320,000 and in 1961 to more than £ 2 million.

Table 17. *Development of Agricultural Exports* in million £

Year	Crops	Animal Products			Total Agric. Exports	Agric. Exports in % of total Exports
		Hides/ Skins	cattle/meat products	Total animal products		
1950	20,23	1,38	·	·	·	·
1951	36,31	1,67	0,32	1,99	38,30	97
1952	38,69	1,05	0,84	1,89	40,58	87
1953	28,70	1,65	0,93	2,58	31,28	91
1954	27,91	1,53	0,83	2,36	30,27	84
1955	27,80	1,23	0,49	1,72	29,52	81
1956	35,69	1,20	0,33	1,53	37,22	83
1957	30,42	1,22	0,51	1,73	32,15	82
1958	30,94	1,20	0,91	2,11	33,05	80
1959	32,69	1,92	1,51	3,43	36,12	80
1960	40,98	1,83	1,94	3,77	44,75	82
1961	·	1,76	2,05	3,81	·	·
1962	·	1,42	2,32	3,74	·	·

Sources: a) Annual Reports, Department of Agriculture.
b) Quarterly Economic and Statistical Bulletin, E.A.H.C.
c) Budget Surveys.

Table 18 contains quantities of various export products. Sisal, cotton, coffee, cashew nuts, and tea account for 85 per cent of agricultural exports (animal products excluded) and 63 per cent of total exports. The remaining nineteen export items are of less significance. Quantities are small, and fluctuations from year to year high. A trend toward expansion is evident for sesame, dried manioc, groundnuts, castor seed, and perhaps soya beans. The maize export for 1960 is a chance result of the especially good harvest in that year. The export of beans, sunflower seed, papain, tobacco, kapok, copra, millet, sorghum, and rice does not show any noticeable tendency towards expansion.

Tanganyika's export gains were unexpected. In 1955 the East African Railways and Harbours requested an estimate of the probable developments in commerce, transport and consequently in agricultural exports, projected up to the year 1974. One section of the resulting study concerns Tanga-

Table 18. *Export of Agricultural Products* [1,2] in 1000 tons

Crop	1950	1951	1952	1953	1954	1955	1956	1957	1958	1959	1960	1961	1962
sisal	119	142	158	171	168	174	186	192	198	209	207	207	220
cotton lint . . .	7,0	8,3	11,1	14,8	12,1	20,4	27,9	27,2	31,1	30,7	38,9	29,7	32,6
coffee	14,9	16,6	18,6	14,2	19,3	18,4	21,5	18,0	21,6	19,6	25,1	24,6	25,7
cashew nuts . .	6,5	8,2	11,5	11,4	16,3	18,2	16,7	33,7	31,3	33,2	36,7	40,0	49,0
tea	0,5	0,8	1,0	1,1	1,6	1,7	2,0	2,2	2,3	2,7	3,2	3,2	3,9
groundnuts . .	0,1	3,5	9,4	1,1	2,5	5,6	15,1	16,1	12,6	12,1	14,6	3,4	6,4
castor seed . .	3,3	7,6	12,3	11,8	9,6	7,9	12,0	13,9	17,9	17,2	18,4	10,7	13,4
beans/peas . .	8,1	10,6	23,2	10,6	10,4	8,3	9,4	5,4	9,5	11,4	11,5	7,2	10,8
maize	.	0,5	8,3	0,2	.	35,8	106,4	9,1	1,2	6,2	43,7	2,5	0,1
sesame	.	.	0,5	0,3	3,1	4,2	10,6	8,2	8,5	11,2	10,8	11,5	8,2
sunflower seed .	8,9	9,5	17,7	6,2	10,2	12,5	14,7	19,6	11,6	5,7	12,5	10,7	11,8
manioc (dried) .	0,1	6,3	15,2	0,6	1,7	2,5	24,0	20,1	17,8	13,7	12,4	.	23,1
gummi arabicum	2,3	1,7	0,6	1,3	1,5	2,1	1,5	0,7	0,5	1,8	1,4	.	.
cotton seed . .	.	.	.	.	.	.	.	8,4	7,5	5,0	5,5	.	.
papain	0,02	0,02	0,04	0,08	0,1	0,04	0,02	0,02	0,07	0,07	0,07	0,07	0,05
tobacco	1,1	0,8	0,6	0,3	0,4	0,4	0,1	0,3	0,2	0,3	0,7	0,5	0,4
kapok	0,7	0,4	0,4	0,8	0,6	0,9	0,7	0,9	0,9	0,9	0,9	0,8	0,8
copra	0,9	.	5,5	1,1	2,0	4,7	3,8	0,2	2,3	3,9	1,0	.	.
millet/sorghum	.	4,1	2,8	1,9	.	13,1	5,1	0,9	2,6	1,8	3,4	.	.
soya	.	.	.	0,7	0,5	0,6	0,9	1,5	0,7	1,0	1,3	.	.
palm kernels .	0,3	0,3	0,3	0,6	0,7	0,4	0,5	0,5	0,7	0,9	0,8	.	.
rice	.	0,2	0,2	.	0,1	1,2	1,7	1,9	0,9	0,1	1,5	.	.
other products .	9,8	2,1	1,2	4,5	3,8	5,6	3,9	4,7	2,8	6,5	9,2	.	.
Total	184	224	298	255	265	339	465	376	383	395	461	.	.

[1] Including deliveries inside the East African Market.

[2] The figures given in the Annual Reports for export volumes are not always consistent. This table contains the most recent figures for each product.

Source: Compiled from a) Annual Report, Dept. of Agriculture.
b) IBRD: The Economic Development of Tanganyika, Table 3, p. 13.
c) Budget Survey 1962—63, Table 14, p. 14.
d) Statistical Abstracts 1961.

nyika[1]. There it is assumed that by 1974 sisal exports will amount to only 200,000 tons. In 1962 an export volume of 220,000 tons had already been achieved. For tea, exports of 2,000 tons were estimated for 1974, in contrast to the figure of almost 4,000 tons in 1962. The export volume of oilseeds and nuts in 1960 was already twice as great as the projected figure for 1974. As to cotton and pyrethrum the estimates for 1974 were reached in 1960.

Table 19. *Net Import of some Important Agricultural Products in 1000 £, (international trade excluded)*

Produce	main supplier	1955	1956	1957	1958	1959	1960	1961
dairy products .	Netherlands	405[1]	370	464	446	456	494	648
rice	Thailand	45	57	208	67	108	52	12
sugar . . .	Taiwan	979	1024	837	491	842	836	805
wheat . . .	Australia	n.a.	97	164	234	123	55	79
maize[2] . .	USA	—	—	—	9	2	—	1210
onions. . .	Syria/Lebanon	32	42	21	41	27	53	45

[1] Net Home Consumption.
[2] Grant from the USA to alleviate the 1961 famine.

Source: Annual Trade Reports, E.A.C.S.O. (or E.A.H.C.).

On the other hand, there are substantial agricultural imports. Table 19 contains data for agricultural imports from countries outside the East African Common Market. Milk products — mainly powdered milk — are imported principally from the Netherlands. Thailand supplies rice, Taiwan sugar, Australia wheat and Syria and Lebanon onions. The heavy import of maize in 1961 was a grant from the USA which was shipped freight-free to Dar-es-Salaam in order to prevent famine. The harvest in 1961 was extremely poor.

III. Inter-territorial Agricultural Trade

It must be remembered that export and import figures for Tanganyika usually concern commerce with countries outside the East African Common Market. In addition, some trade takes place with Kenya and Uganda. Tanganyika imports the following items from Kenya: wheat, wheat flour, tea, fruit, and vegetables, both fresh and preserved, meat and meat products, fresh milk and cheese. The most important import from Uganda is cotton-seed oil (Table 20). Processed agricultural products such as beer and cigarettes are also imported by Tanganyika.

[1] East African Railways and Harbours: A Study in Trends: The Economy of East Africa. London 1955, p. 123.

Tanganyika supplies Kenya with onions, leguminosae, raw tobacco and dried pyrethrum flowers [1]. Uganda obtains small quantities of rice, maize, millet, and ghee from the territories around Lake Victoria. In some respects Tanganyika is in the unfavourable position of a supplier of raw material. Processing is done mainly in Kenya. Tanganyika is the main producer of

Table 20. *Inter-territorial Commerce in some Important Agricultural Products* in 1,000 £

Product	main supplier	1955	1956	1957	1958	1959	1960	1961
I. Tanganyika's net imports from Kenya and Uganda								
Wheat	Kenya	·	·	62	13	398	638	625
Wheat flour . . .	Kenya	493	378	505	565	317	282	204
Tea	Kenya	57	193	386	371	393	305	434
Fruit/vegetables [1] .	Kenya	48	65	81	116	147	170	239
Meat/meat products	Kenya	69	98	124	80	70	112	216
Milk/milk products [2]	Kenya	139	125	116	200	145	185	164
cotton-seed oil . .	Kenya	116	95	189	164	255	343	354
Sugar	Kenya	(—25)	(—28)	314	360	114	62	·
II. Tanganyika's net exports to Kenya and Uganda								
Onions	Kenya	·	·	55	111	104	146	181
Leguminosae . . .	Kenya	221	178	137	171	104	218	157
Rice	Uganda	44	77	34	76	53	72	90
Maize/maize flour millet . . .	Uganda	·	·	71	257	257	81	26
Tobacco, unprocessed . .	Kenya	329	268	451	440	246	344	397
Pyrethrum leaves .	Kenya	52	82	64	71	99	86	95
Ghee [3]	Uganda	19	117	126	92	74	66	56

[1] Fresh and preserved, onions excluded.
[2] Ghee excluded.
[3] Melted Butter. Figures for 1955—58 indicate gross export.

Source: Economic and Statistical Review (or Bulletin), E.A.C.S.O. (or E.A.H.C.).

raw tobacco in East Africa. Part of Tanganyika's cigarette consumption is manufactured in Kenya. Tanganyika is a great exporter of coffee and tea. Domestic demand is almost entirely covered from production processed and packaged in Kenya [2].

[1] In 1962 a pyrethrum factory which will process Tanganyika's crop, was opened in Arusha.

[2] Concerning the problems to which this situation gives rise, see the Raisman report.

Chapter D

Agricultural Development Policy under British Administration

I. Development Policy until 1950

Economic activity characterized the final decade of British administration more so than its earlier years. The change in political status which followed World War I undoubtedly had negative effects on Tanganyika's agricultural development, both in the sphere of the estate economy and of peasant farms. In the last decade of the German colonial administration the viewpoint had gained increasing support that estates alone could not achieve the desired production, that the conditions were unsuitable for settling German farmers on family holdings and that therefore African farms were to be encouraged. Since the turn of the century [1] the influential Cotton Committee, for example, supported the introduction of cotton cultivations as "a peoples crop". A report from as early as 1907 states:

> "The cultivation of cotton by the Africans ... is experiencing continuous development. In many parts of the colony the Africans have already recognized the advantages of this crop, and one may assume that with further encouragement cotton planting will become a permanent practice."

In Sukumaland cotton seed was distributed free of charge. Premiums were paid for the planting of cotton and a textbook on cotton cultivation was published in Swahili. The establishment of a cotton research centre close to Mwanza was scheduled for 1914. In 1913 there were 32,000 acres planted with cotton on estates, and 47,000 acres on African farms.

British administration between the two World Wars encouraged commercial production on African farms. The successful development of peasant-coffee on the slopes of Mt. Kilimanjaro is well known. Yet the overall emphasis was placed less on agricultural development than — in

---

[1] At that time Germany imported cotton from overseas, valued at 420 million marks, and three-fourths of this was supplied by the USA. It was hoped that the introduction of cotton cultivation in the German colonies might influence price formation. In this case even relatively expensively produced cotton might have been commercially profitable (see SCHANTZ-CHEMNITZ, „Das erste Vierteljahrhundert deutscher Kolonialherrschaft", Sonderdruck, Cöthen, 1909 [?]). The German administration also supported coffee planting by Africans. Coffee farmers were exempted from performing otherwise compulsory work. In Moshi-District, in 1916 approx. 16,000 coffee trees were owned by African farmers. (Agricultural District Book, Moshi.)

accordance with the provisions of the mandate — on orderly administration. There was very little money and personnel available for development purposes. Tanganyika is economically less lucrative than neighbouring Kenya. It was a mandated territory and not a crown colony, i.e. it had a less certain political status. Great Britain, like Germany before her, was not especially interested in obtaining raw materials. Thus Tanganyika, due to the logic of circumstances, held the position of stepchild among the British East African territories.

Corresponding to the ideas of the times and the tradition of the Colonial Office, attention was directed toward the protection of African interests, but in a static sense rather than in the sense of accelerating economic and cultural change. This emphasis on the one hand effected a delay in economic development, but on the other made possible the development of a relatively peaceful political atmosphere. Comparatively few long-term leases were granted to Europeans and Asians.

The situation after World War II was different. Efforts were made towards economic development and development plans were being worked out. However, little importance was attached to commercial farming by Africans. Presumably the idea that the African farmers could become the mainspring of the economy was still inconceivable. More rights of occupancy were given to Europeans. Lord Delamere spoke of a "white belt" in the African highlands as early as the 1930's. This idea was again taken up after World War II. The settlement of numerous British farmers in the Southern Highlands of Tanganyika was to provide a link between the European Settlements in Kenya, Rhodesia and South Africa [1]. Other factors were also considered. More commercial production was needed, in order to obtain greater tax revenues with which to finance additional social services for the African population, i.e. education, administration, roads, employment, etc. The way to this was seen in more estate farming.

The most extensive project of the time, the groundnut scheme, was, however, motivated by other considerations. Three million acres of bush, mainly in Tanganyika (Kongwa, Nachingwea and Urambo) were to be cleared by tractors and sown with groundnuts. The production of groundnuts was to alleviate the world shortage of fats. In fact, approximately 220,000 acres were cleared in 1947—49. In 1949 the groundnut scheme was discontinued as suddenly as it had been started. It cost £ 35.87 million and left a cleared area of land which required still more money to make it arable. A successor organization, the Tanganyika Agricultural Corporation (TAC) took over management of the cleared land with a new balance-sheet.

---

[1] The attempts to settle greater numbers of European farmers near Iringa (Southern Highlands) failed because of faulty organization, unsuitable settlers and poor soil.

Many mistakes are attributed to the groundnut scheme: too rash a beginning, too sudden an end, transfer of commercial activities to ex-Army-men, poor purchasing, false planning, waste, etc. Most positions were occupied by personnel who had neither experience of the tropics nor of farming. Although is was evident after the first year that the project was heading for a disastrous failure, it was continued. Plant diseases and drought ruined the harvest when the first large production was expected. This explains the high costs. Yet the main reason for the complete failure of the scheme is that mechanized clearing and motorized farming of large areas of marginal soil does not pay. Groundnuts do grow in the areas under discussion. Yet yields of 400—800 lbs/acre are not high enough to cover a high monetary outlay. For the last ten years TAC has been trying to manage what remains of the groundnut scheme on commercial principles. The result as to crop farming: Large farms in Nachingwea are hardly able to cover costs even today, not to mention the amortization of the investment in land clearing. On marginal land, hoe cultivation by the African farmers is apparently more competitive [1].

It would be unfair to attribute the failure of the groundnut scheme to the Agricultural Officers of the British colonial administration [2]. Their approach to development was different. The groundnut scheme was an exceptional case, which was decided on in London. The politicians of the Labour Government of the time apparently had unrealistic ideas about the advantages and disadvantages of large agricultural projects in tropical Africa.

The remarkable increase in production between 1950 and 1960 is thus not the outcome of big schemes, or — as in Kenya — the result of additional white settlement. It is mainly due (1) to continued development in the already existing estate economy and (2) to additional peasant production. One has to ask therefore: What kind of agricultural development policy was pursued by the British administration? What measures brought success or failure, and why?

---

[1] The same was true of other trial with motorization in the early fifties. The Agricultural Department established small stations with two or three tractors. They were financed by Local Authorities and run by the Agricultural Department. In the beginning the charge for ploughing was 40 shs/acre. Some schemes had a good start. In the Rufigi area 7,720 acres were ploughed in 1952. Yet costs were higher than expected. Peasant fields are small and not consolidated. Finding and measuring small plots of land required much time. It was extremely difficult to get the farmer to pay for services rendered. None of the projects covered costs. They amounted to an average of 76 shs/acre. When the charge was fixed at 60 shs/acre, peasants stopped asking for tractor ploughing. All of the projects had to be abandoned.

[2] The idea originated from a member of the commission which prepared the report. He was, however, the director of the Agricultural Department in Dar-es-Salaam.

II. The Early 1950's — Administrative Ordinances as an Instrument of Development

1. Additional Development Services

Among the numerous factors which contributed to the remarkable production increases in peasant-farming, two are especially evident:

— the high prices of the Korea boom aroused peasants from their usual lethargy and gave them an initiative to produce for the market;

— The Department of Agriculture received more personnel and more funds. The money was used mainly for expanding general agricultural services and for numerous small projects, mainly to improve peasant farms.

The Ministry of Agriculture was primarily responsible for providing three services [1]:

— The *Agricultural Department* was responsible for the organization of local markets, for providing aid in cases of famine and flood, improving production by supplying better seeds, introducing new crops, popularizing better methods of land use, advising the Local Authorities [2] on irrigation, land reclamation, running of nurseries, controlling agricultural credit, carrying out agricultural research and education etc.,

— The *Veterinary Department* also had extensive responsibilities: cattle markets, slaughter houses, meat inspection, leather industry, animal health, including control of the tsetse-fly, animal breeding, and animal husbandry [3].

— The work of the *Department of Water Development and Irrigation* consisted mainly of providing drinking water for man and animals by the construction of small dams and wells. Some of the dams include small irrigation projects. Upon completion the management of these projects was transferred to the Local Authorities. The projects were financed mainly by government credits which had to be repaid with interest by the Local Authorities from tax revenues.

---

[1] The names and responsibilities of the Ministries changed from time to time. The Ministry for "Natural Resources", later known as the Ministry for "Agriculture and Co-operative Development" and since 1962 as the "Ministry of Agriculture", was responsible for various "Departments", later "Divisions": "Agriculture, Veterinary Services, Fisheries, Forests, Water Development and Irrigation" and sometimes also for "Co-operative Development" etc. In 1962 the responsibilities of the Ministry of Agriculture were limited to arable farming, animal husbandry, water development and irrigation.

[2] Formerly known as "Native Authorities". They were concerned with African administration on a local level.

[3] Responsibility for cattle-markets passed into the hands of the Local Authorities. Since 1962 the Agricultural Department has been responsible for animal husbandry.

The work of the Agricultural Department has been of predominant importance in agricultural development. The staff was greatly increased. The number of officers with higher education rose from 127 in 1950 to 232 in 1960 (Table 21). In 1950 the work was carried out by 756 instructors who had hardly any training, in 1960 by 815 assistant field officers, most of whom had completed the two-year course at the agricultural school in Tenderu (Northern Region) and Ukiriguru (Lake Region). In addition, 1,210 instructors without formal training were employed in 1960.

Table 21. *Staff of the Agricultural Department 1950—1960* [1–8]
Number of employed

Year	Agricultural Officers	Specialists	Field Officers	Field Assistants	Agricultural Instructors
1950	39	17	71	·	756*
1951	46	21	79	·	·
1952	52	23	91	·	971*
1953	48	24	101	·	1512
1954	47	21	91	414*	1442
1955	50	20	97	527*	1264
1956	49	20	103	322	920
1957	66	25	112	411	872
1958	70	28	114	608	1246
1959	69	30	138	700	1272
1960	74	33	125	815	1210

[1] Compiled from Information given in the Annual Reports. The figures marked with * are taken from the Report of the World Bank, Table 16, p. 61. The figures from the latter report do not correspond exactly with those given in the Annual Reports.

[2] Including personnel for fishery services, excluding personnel of the Forestry and Veterinary Department.

[3] The figures apply to personnel employed as at December 31st.

[4] Part of the personnel is always on vacation, part on military service.

[5] The figures also include personnel in Dar-es-Salaam.

[6] In most cases the Agricultural officers, specialists and field officers have completed courses at an agricultural college. In 1960 the first senior field assistant was promoted to field officer.

[7] At first the field assistants were former agricultural instructors who had been promoted to these posts, more recently many of them have completed the two-year course at the agricultural schools at Tengeru and Ukiriguru. The name has been changed to assistant field officer.

[8] No new agricultural instructors are being employed.

2. Improvement through Compulsion

The main problem of agricultural development in a country such as Tanganyika is the uninterested attitude of the peasants towards economic efforts and rewards. They have a rather high appreciation of leisure. They like to have money and consumer goods, but they shy away from the effort necessary to obtain them. Moreover they think in static terms, i.e.

they accept the existing order of things as the natural one. BLOHM reports on the Nyamwezi (and this statement, 20 years ago could apply to most of Tanganyika), that:

> "... their lives are controlled by laws which might even be considered stringent. He who violates these inherited and unwritten laws is not a person of 'repute'. If, out of ignorance or laziness, a person violates these laws, he is considered a despicable fool; if a person ignores them because he is stubborn or feels he knows better than the rest of the community, he is probably a 'mudaki', i.e. in this connection a person who does not listen to the good advice of his neighbours, or he is possibly even a 'mulogi', i.e. a witch doctor." (Part II, p. 65.)

In contrast to this attitude, agricultural development is characterized by the principle of change. That a rapid population increase calls for a rapid change in the economic structure has not, until now, penetrated the minds of rural people. The "old ones" who, according to TEMPELS "Bantu-Philosophie" are also the "wise ones", are particulary critical of new ideas [1]. They consider themselves responsible for protecting the customs which have been handed down to them. For example: the local chief demanded that the first tin roof in Sukumaland be torn down. The first master farmer in the Usambara was murdered. According to the local Agricultural Officer, the first cotton seeds introduced in the Pare Mts. were cooked before being sown, in order to prove that such a crop could not be grown in this area. The tenacity with which peasants in Sukumaland cling to their old habits is evidenced by the fact that even after six years of working on experimental fields where yields of 1,000 lbs/acre of seed cotton were obtained, the very persons who did the work there still cultivate their own fields in the usual manner, producing yields of only 400 lbs./acre.

Under British colonial administration, officers of the Agricultural Department tried to counter resistance to agricultural progress through administrative ordinances. N. V. ROUNCE, one of the most remarkable and most successful Agricultural Officers in Tanganyika, wrote, in the spirit of the times and of his Department: "the African will have to be compelled to help himself". Administration ordinances were considered the proper instrument to get peasant development. Local Authorities were instructed to issue agricultural regulations. The implementation of such regulations was supervised by officers of the Agricultural Department. Cases of non-observance were reported and punished by the Local Authorities. The increasing activity of the Department of Agriculture consequently meant more ordinances and more controls. Following are three examples of the procedure which was used.

[1] TEMPELS, B.: Bantu-Philosophie, Heidelberg, 1956.

Manioc in the Tanga Region. A regulation requiring the cultivation of manioc on a specific area of land was already effective during the German colonial administration. Manioc was to serve as a protection against famine in case of poor maize or millet harvests. This regulation was revived in the late forties. The coastal strip of the Tanga Region had become an area of chronic hunger, even though sufficient land was available for manioc cultivation. A change in the nutritional situation was brought about by forced cultivation. The farmer who did not plant was fined. Agricultural Officers excercized control. They often had to be accompanied by armed policemen when carrying out their duty. In the case of manioc cultivation in the Tanga Region success was achieved. Today a surplus is produced, dried and exported.

Coffee on Mt. Kilimanjaro. Peasant coffee on Mt. Kilimanjaro was, and is today, not as well cared for as it should be. Administrative ordinances were used to achieve improvements. The Chagga Council ordered, for example:

— Only planting material which has been certified by the Agricultural Department may be used;
— Coffee may be planted only after the field has been examined by the Agricultural Officer or his representative, and found suitable for this purpose;
— Agreement to the planting of coffee will be granted only after proper preparation of the land and the provision of sufficient shade;
— Those persons who ignore these regulations will be punished by having their coffee trees torn up, and will have to carry the costs of this measure.

The work of the Agricultural Officers consisted mainly of supervising instructors who were to implement these measures. Because of a shortage of personnel in the Agricultural Department a thorough implementation of the directives was not possible.

Iringa Dipping Scheme. One of the most prevalent cattle diseases in Tanganyika is East Coast Fever. This disease, which is carried by ticks, is responsible together with the tsetse-fly, for preventing the use of large area of grazing. Cattle is concentrated in healthier regions which are consequently threatened with overgrazing and erosion. East Coast Fever can be controlled. Animals have to pass through a "dip" at regular intervals, which contains a chemical fatal to ticks. There are always some cattle owners, who dip. The difficulty lies in persuading all of them to participate, especially since a small fee is charged. In Iringa District grazing was sufficient to justify an expansion in the number of cattle. In 1953 the Iringa Dipping Scheme was begun. If successful it was to be followed by other such schemes.

A specific area was included in the project. Veterinarians explained the aims and the advantages of the project to local chiefs. Twenty supervisors organized the participation of all livestock in the area. Dipping facilities had been built in the most suitable locations.

The technical success was impressive. Within four years East Coast Fever was almost completely wiped out. The number of cattle rose from 161,000 to 190,000. In spite of this, the project met with resistance. Cattle owners did not ask for the services rendered. It had been forced upon them. Chiefs who co-operated were obviously no longer the "voice of the people". Social change had already weakened their position. Ordinances, requiring the herdsmen to bring their cattle again and again (and without apparent end) to distant dips, for which in addition fees had to be paid, were considered chicanery. Local politicians, who understood the purpose of these measures as little as did the cattle-owners, took advantage of the latter's discontent. The project could not be maintained in face of the resistance of the people, especially since Tanganyika was under UN-trusteeship. Several years later most of the dips had simply disappeared. They had been dismantled and what was sellable was sold.

One must not use the example of the East Coast Fever Project as an indication of the failure of veterinary work in Tanganyika. Much work was done with remarkable successes. In 1956, for the first time, no cases were recorded of Rinderpest. The danger remains, however, as the disease can be contracted through game. The threatened areas are, however, encircled by a belt of obligatorily vaccinated cattle. Due to much veterinary work it is meanwhile technically possible to control economically significant cattle diseases in Tanganyika. Such control is also economically profitable as long as it is carried out within the framework of a rational system of animal husbandry. This is also true for the tsetse. An example is Amboni Ranch, near Tanga. The problem lies in the practical application of proven veterinary measures. The failure of the Iringa dipping scheme indicates that force — at least when applied under the political conditions of a weakened colonial administration — is not sufficient to carry out measures which are useful to cattle owners.

3. The Priority of Erosion Control

Another remarkable characteristic of the early fifties is the priority given to erosion control. Efforts aimed not so much at increasing commercial production. They focused on the conservation of natural resources. Increasing erosion was recorded with concern. In accordance with the spirit of the times, the belief was held that "the health and the vigour of a nation depends upon the health and vigour of its soil". Ordinances and control were considered to be the proper way to fight erosion. Following are three examples:

Erosion Control in the Usambara Mts. In 1951 it was announced that the following ordinances would gradually be made compulsory in various areas of West Usambara Mt.:

- Cattle may only be kept in stables.
- Cattle may be grazed only on flat land.
- Slopes exceeding 25 per cent may not be cultivated.
- Slopes of less than 25 per cent may be used only with ridging.
- Weeds and crop refuse may not be burned. They have to be used for trash bunding.
- Elephant grass (penisetum purpureum) is to be planted as a fodder crop.
- Cow manure from "bomas" is to be transported to the fields.
- A 5 yard strip of land planted with elephant grass, sugar cane or bananas must run on both sides of all watercourses.
- Irrigation is allowed only on flat land or in conjunction with ridging.
- Steep or eroded slopes are to be planted with trees.

The Usambara project was not based only on ordinances and controls. Extension Officers were concerned with irrigation, vegetable farming, poultry etc. Resettlement on irrigation projects in lower valleys was planned. Ordinances concerning erosion control were, however, the heart of the project. Where necessary, implementation was forced upon peasants by administrative measures. In 1954 a total area of 45,000 acres was ridged or terraced. Yet farmers objected to the additional work. In 1957 growing unrest led to the suspension of these regulations. In some cases, progress which had been achieved was intentionally destroyed. Today one can only occassionally recognize the remainders of earlier ridging or terracing. The only improvement which lasted was the financially advantageous cultivation of bananas on steep slopes. For the rest, the Usambara scheme was a total failure.

Erosion Control on Grazing Areas. One of the most important cause of erosion is to be found in the anachronistic organization of animal husbandry. In various areas efforts were made to rationalize the use of grazing via specific grazing schemes. (Gogoland, near Iringa, in the Mbulu District, near Nzega, and near Kondoa.)

Here is an example from Kondoa: The number of cattle which could be fed in a certain area without causing erosion was determined. Every livestock-owner received "coupons" which entitled him to graze a certain number of cattle in the given area. The excess cattle were to be sold or were allowed to migrate to areas which had been opened for grazing by additional watering-places. Coupons could be traded.

The watering-places which were built under this project are still in use. They have contributed to an increase in stock and consequently to increasing erosion. The most important element, i.e. the control of cattle numbers, could not be carried out anywhere for more than a few years. Almost all grazing schemes ended in total failure.

Erosion Control in the Uluguru Mts. The tributaries of the Ruvu river, which is important for the water supply of Dar-es-Salaam, flow from the Uluguru Mts. near Morogoro. Studies by SAVILE indicated that water supply was endangered by the soil and forest-destroying practices of mountain peasants. A scheme was begun in 1949 which was to (a) improve land use, (b) preserve and expand forests, and (c) stabilize the flow of water. £ 68,000, five Agricultural Officers, and sixty instructors were employed for this purpose. The scheme started in three pilot areas of one sq. mile each. They promised to be successful. Personnel then toured the villages. For six months the scheme was explained and the advantages pointed out.

Then, however, obligatory labour was required. Farmers had to terrace their fields. Work was supervised by instructors. YOUNG & FOSBROOKE give the norm as 300 yds. of terraces per year and household. The project gained momentum. 900,000 yds. of terracing were done from 1950 to 1952. From January to October 1954 more than 11 million yds. had been terraced. In the periodical reports some difficulties were spotlighted. Altogether these reports gave the impression of successful work, as did most reports of this kind in Tanganyika.

As time passed, however, farmers developed growing resistance. They were convinced that terracing reduced the yields — a contention not totally unjustified from a short-term point of view. Consequently they terraced first the most infertile slopes. They complained that the instructors treated them as schoolboys. They could not understand why they should work intensively 2—3 days per week for something which reduced their incomes. They tried to bribe the instructors, so that they could escape work and avoid reduced yields. Burning of crop refuse was one of their oldest customs. Under this project burning was forbidden. Crop refuse had to be used for trashbunding. Farmers resisted, again not without cause, since maize borers breed in maize straw and damage the next crop. The hoe proved to be an unsuitable tool for terracing. Spades would have been better, but since peasants have no shoes, they would have had to use their bare soles to operate normal spades. Special spades for bare-footers were not available.

Discontent grew into open riot with the spread of rumours that forced resettlement from the mountains to the flat land was planned. The cry was raised "we want the customs of the old". Politicians organized discontented farmers by stating that an African government would restore "the ease of the old life". Instructors were attacked physically. In Morogoro protest demonstrations were held. The threatening behaviour of the crowd forced the administration to resort to arms. One death resulted. Peasants refused to continue with terracing. Terraces which had already been completed were wilfully destroyed. Throughout the land bush fires burned as a symbol of protest. 700 men came to help those who had been arrested and brought to court for inciting riot. They were armed with clubs and knives and besieged the court-room, effectively placing pressure on the court of the Local Authorities.

The Uluguru scheme failed. Only a few of the terraces, which had been built with so much effort, are still in existence. Deforestation continues. Efforts to control erosion resulted in the opposite of what was desired. At present, erosion control in the area is more difficult than it has ever been. The reason for the failure was the direct approach to the objective of erosion control. Farmers are disinterested. It would probably have been better to start a scheme like this with the accent on improving the cash income. In due course soil conservation measures can be coupled with the introduction of profitable innovations. The cultivation of cash crops which are at the same time tree or bush crops. i.e. coffee or bananas, provide protection for the soil. If the cultivation of coffee is something new, it can be introduced together with mulching, terraces, ridges, grass planting as a "package deal", and will thus be accepted into local usage.

The damages of the Uluguru Project are not limited to the loss of money and labour. The Agricultural Officers and the instructors were degraded to "policemen". In the following years extension officers faced a wall of mistrust. The Annual Report of the Agricultural Department for 1961 states therefore: "The extreme conservationist is as dangerous as the land miner." (I. p. 33.)

4. Increased Cash Crop Production: Cotton in Sukumaland

Another scheme, however, the largest one, achieved remarkable success. In 1947 a grant of £ 520,000 was given for the development of Sukumaland [1]. Within ten years the following objectives were to be achieved:

— A stop to the loss of soil fertility in the cotton-growing area around Lake Victoria,
— Increasing yields per acre,
— Introduction of a stable system of farming which preserves soil fertility,
— Expansion of arable cultivation, especially in Geita District.

The first three objectives were not achieved. Much more land was cultivated. Better land use, however, was not realized. The administration tried to obtain planned land reclamation via ordinances. Population density and livestock numbers were to be controlled, forest reserves on hills were to be conserved and areas threatened with erosion were not to be cultivated. All this was not achieved. Today's farming in Geita District is as much a case of soil-mining as in other parts of Sukumaland.

The expansion and improvement of cotton cultivation brought unexpected success, however. Money-making via cotton became the central point

[1] Main area of the Lake Region. Figures concern the region.

of the whole design. In 1950 the cotton crop in the Lake-Region amounted to 40,000 bales. The 1963 crop is estimated to exceed 220,000 bales. Production increased almost sixfold in ten years. Cotton thus became the second most important export product, taking its place between sisal and coffee. Practically the entire production is produced in Sukumaland which, in the

Table 22. *Development of Cotton Production in Lake Region*

Year	Total Production in 1000 Bales (lint)	Price cts./lb.[1]	Acreage[2] in 1000	Number of cotton growers in 1000	Acreage/ Grower	Yield/[2] lbs/acre	Percentage of crop handled by coops.
1950	40	34	·	·	·	260	·
1951	40	50	·	·	·	275	·
1951	70	50	166	·	·	420	·
1953	39	50	90	·	·	420	13
1954	91	62	216	159	1.36	420	32
1955	109	62	·	·	·	·	44
1956	121	57	336	219	1.54	421	63
1957	151	54	·	268	1.47	474	70
1958	152	54	400	248	1.61	460	83
1959	183	52	420	·	·	450	100
1960	161	54	·	·	·	·	100
1961	160	54	·	·	·	·	100
1962*	195	55	·	·	·	·	100
1963*	225—260	51	·	·	·	·	100

[1] Producer price for seed-cotton.
[2] The figures indicating acreage and yields are rough estimates.
* Provisional figures.

Source: Compiled from
a) Annual Reports, Department of Agriculture.
b) Report of the World Bank, pp. 12, 13, and 202.
c) Empire Cotton Growing Corporation: Progress and Reports from Experiment Stations, Tanganyika, Lake Region.

process, became an important economic and political center of the country. This is an indigenous economic achievement, and therefore deserves double recognition in the current African situation.

The fact that this was a success raises the question of the reasons why. In conversations with participants and observers of the scheme the following points were made:

The Opportunity to earn Money. The emphasis was placed not so much on conserving soil fertility, as formulated in the objectives, but on making cash through increased cotton sales [1]. This appeal was systematically supported. Traders

[1] F. Stuhlmann reports. "The Wasukuma are interested in progress. They love European goods, in particular textiles. They are skillful workers and, partly, acceptable soldiers." (See: Mit Emin Pascha ins Herz von Afrika. Berlin 1894.)

in Sukumaland were encouraged to display such sought-after commodities as textiles, bicycles, tin roofs, etc. and to keep them in sufficient supply.

The price of cotton rose significantly, due to the Korea boom. In 1950 producers received 34 cts./lb of seed cotton, in 1951 the price was 50 cts. and in 1954 62 cts. This price increase was decisive. Theoretical considerations would seem to indicate that African peasants who have hardly any cash costs for crop cultivation and hardly any other alternative for their labour and that of their family, except to till the fields, are better off if they produce more, even when prices are low. One could perhaps expect a negative price elasticity of production, since with increasing prices the same income can be obtained with less production, i.e. less effort. Events in Sukumaland indicate the opposite. The high cotton price persuaded farmers to produce for sale. Once they had become accustomed to a certain consumption level, falling prices led them to expand production in order to preserve recently gained living standards. In any case, the value of cotton production in Sukumaland has doubled from 1955 to 1959, though the returns per lb decreased by 10 cts. or approximately 20 per cent.

Another incentive was the desire to own more cattle. The herds were reduced by disease and drought. In 1950 the Lake Region had only half the cattle reported for 1962. Peasants were interested in cotton, in order to obtain money, save it, and buy cattle and thus to gain in security and social status.

Agricultural Administration and Extension. The Sukumaland Scheme was a long-term one for which comparatively large sums of money were made available. The Agricultural Department contributed the largest possible share of its personnel[1]. In addition local instructors were recruited by Local Authorities who, although responsible for the scheme, delegated most of the supervision to the officers of the Agricultural Department. In 1952 1,286 officers and instructors were employed. The development policy was shaped by N. V. Rounce. He had known the Wasukuma for years, spoke their language and was familiar with their customs. The personnel at his disposal were men who wished to make a career in colonial service. He supervised them with unusually stiff discipline. It is open to discussion whether it was wise to employ large numbers of untrained instructors. They performed their work poorly. Farmers were more ordered about than advised. Some instructors accepted bribes. Farmers called them "black locusts". On the other hand one has to consider that with the instructor there was somebody in the village focussing attention on the subject of cotton. The very fact that there was much talk about cotton might have contributed to the growing interest in cash crop cultivation.

The scheme made use of several decades of scientific research. Since the beginning of the thirties experimental work at Ukiriguru had concentrated on cotton. The work was done in co-operation with the Empire Cotton Growing Corporation which had world-wide experience in cotton. Thorougly tested improvements were available. Plant-breeding was especially effective in Sukumaland, as it usually is in cases of spectacular agricultural success. New and undoubtedly better cotton

---

[1] The institutional regulations altered from time to time. Local Authorities were responsible for the Sukumaland Development Scheme in its initial stages in 1947. Later the Lake Province Cotton Committee, which was financed by the Lint & Seed Marketing Board, was founded. In 1955 the South-East Lake Cotton Council was given responsibility. The council was financed principally by the Lint & Seed Marketing Board. The agricultural specialists, with the exception of a few permanent employees, were seconded by the Agricultural Department.

seed, developed in Ukiriguru, was distributed. Advantage was taken of the fact that cotton must be processed industrially. Farmers can hardly produce their own seeds. Seed and fibre are separated in the cotton gin. The Agricultural Department, co-operatives and the Lint & Seed Marketing Board made use of this fact. Only certified cotton seed which had been treated with insecticides was released from the ginnery. In addition, cotton was not a new crop. Farmers had been familiar with it since the German times. They knew about various soils and local climate. Established marketing channels were available. The Lint & Seed Marketing Board had at its disposal a well experienced apparatus which organized transportation and sales.

Ordinances and controls played an important role. Local Authorities compiled and distributed a "complete book" of ordinances on land use and cultivation practices. Ridging, manuring, and burning of cotton stalks, was required and controlled. Non-observance was punished by fines. Headmen and instructors often received premiums when large amounts of cotton were planted in their areas, and they therefore put pressure on the farmers. Further pressure came from higher taxes. The obvious way to pay them was to produce and sell more cotton.

The most popular aspect of the scheme was land reclamation. Sukumaland is relatively heavily populated. In the neighbouring western Geita District fertile land lay idle. Land reclamation had not been permitted in Geita before that time. The land belonged to another tribe. Small settlements were threatened by sleeping sickness. The Sukuma Scheme suspended these restrictions. The government built feeder roads and dams, thus making drinking water available. Part of the manual work was done by "tribal turnouts", i.e. unpaid labour performed in the interest of the community. Land reclamation was probably that part of the project with the most favourable input—output relationship. Settlement was inexpensive. Families which flowed into the new area took care of themselves. Their efforts were supported by the not-too-distant home villages. The government's outlay was limited to the construction of a simple infrastructure of roads and water.

The Indians and the Co-operatives. Another factor of great importance was the appropriate organization of cotton sales and processing. Local Indians invested substantial sums in ginneries. In 1950 a ginning capacity of more than 100,000 bales of cotton faced a production of 40,000 bales. Indian ginners were interested in getting more supplies. They helped as much as they could with words and bribes. Efforts by Indian ginners were channelized by administrative measures. Market zones were established and selling points defined. The directive said that every cotton producer should have a selling point within five miles from his home.

Indian and British activity undoubtedly put a spark to African initiative. They established — as a reaction to commercial dominance by foreigners — indigenous cotton co-operatives, thus giving new and complementary incentives to cotton production. Two considerations had stifled the initiative of the Wasukuma farmers. They firmly believed that the Indian traders took advantage of them, that they were not honest in weighing. A few test weighings proved that this belief was not unfounded, though one may assume that the practice of swindling was probably insignificant in the over-all picture. There was strong competition among the Indians. Furthermore the farmers believed that the production of cotton served English interests rather than their own, i.e. that they were working for someone else's benefit. This idea was probably a remnant from the war years in which the Government strove with all possible authority to increase the market supply of essential raw materials.

With the establishment of African primary co-operatives these two psychological handicaps to production were somewhat reduced. P. Bomani, who used cotton co-ops in Uganda as an example, was especially active in organizing the co-operatives in Sukumaland. His example was taken up in the villages. In 1953 13 per cent of the crop was purchased by the co-ops. In 1956 the percentage amounted to two-thirds of the crop. In 1959 the co-operatives obtained a monopoly for the purchase of cotton. After the initial impetus by the Africans, further co-operatives were organized with the support of the Department for Co-operative Development. At one time eight the Co-operative Officers and four Assistant Co-operative Officers were employed by the cotton co-operatives. The individual primary co-operatives were later incorporated into Unions, and later into the Victoria Federation of Co-operative Unions.

A considerable part of the emotional and political unrest of these years was directed towards productive ends by the organization of co-operatives. In Sukumaland the farmers felt that they were working for "themselves". Thus, the animosity against the Indian traders and the colonial administration found a positive outlet. Tribal traditions may also have contributed to the organization of co-operatives and the increases in production.

The Wasukuma are numerically the largest tribe in Tanganyika. They are strongly independent and have been able to absorb small neighbouring tribes into their own group. Their conception of the world is well established and markedly collective. Tribal feelings are strong. Active individual farmers — who are usually the heralds of agricultural development (viz. Chaggaland) — are relatively few in number in Sukumaland. The group or the village, as the case may be, reacts as a unit and this normally makes the introduction of new ideas more difficult. If, however, the group as such accepts a change, a considerable effect can be achieved in an amazingly short period of time. It is possible that just such a collective decision effected the extension of cotton cultivation. Local chiefs played an important role. They were largely members of old — established families of repute. They used their traditional status and their administrative authority in favour of the scheme. Additional animation came from master dancers and master singers at local festivals, who are among the most influential people in Sukuma society.

Difficulties. The Sukuma project did not, however, proceed without difficulties. The policy of issuing ordinances through Local Authorities and the controlling of these regulations by an army of instructors who levied fines for non-observance, gave rise to a growing opposition. The economic success and the increasing awareness of their own achievements fed the opposition of the farmers. In 1958 ordinances — with a few exceptions — were suspended in Sukumaland, as they had been in other parts of the country. It has been reported that at that time some of the farmers collected the manure which they had been forced to transport to their fields, and carried it back to the "bomas". This dissonance occuring after ten years of activity on the project does not lessen the success which has been achieved. The use of ordinances and compulsion was probably indispensable in the first phase of the scheme. One could not do extension work with untrained instructors. In a situation where well trained personnel are not available, ordinances and supervision are the only alternative. The scheme might have been even more successful, had the principle of "persistent persuasion" been applied gradually in the later years. N. V. Rounce died a few years after the initiation of the scheme. Some of his co-workers report that such a change in the methods would have corresponded to his own ideas on the subject.

5. Summary of the Early Fifties

The efforts made towards economic development in the early fifties were not confined to ordinances and control within the above-mentioned schemes. Simultaneously, the less spectacular but — in terms of personnel and effects on production — important routine work of the Agricultural Officers and veterinarians was intensified. Admittedly many schemes failed. It may almost be considered a rule for small agricultural schemes in Tanganyika, that in the beginning some successes are achieved, whereupon optimistic reports are written (at least one was concerned with showing the UN that successful work was being done in developing the country); then, however, obstacles arise and the project is abandoned. The personnel change. After a few years hardly anybody knows that there had been a scheme, which had failed.

Although the number of failures is high, it can be assumed that a comparison of development input and production output will show a favorrable balance. The one great success in Sukumaland is weighing heavier than all other failures put together. Of course, one can ask whether peasants and traders without government would not have achieved similar results. Additional production is undoubtedly not only due to more government outlay. Most peasants hardly ever met an Agricultural Officer.

On this question we are facing a problem to which there is no logical answer. Several factors contributed to increased production. Government effort was one of them. The fact, that additional production and additional effort occur in the same years — while before production was stagnant for decades — is however an indication that government action was successful in touching off peasant initiative, or that peasant initiative was directed through government effort into productive channels.

On the one hand, the number of failures and the growing opposition of the peasants to compulsion required a new approach in agricultural development policy. The balance sheet of experience in the mid-fifties runs as follows:

— The failure of the groundnut scheme, the failure of Local Authorities to motorize farming, and the unprofitableness of the mechanized TAC farms showed that economic conditions were not yet ripe for successful motorization.

— The realization of agricultural improvement by administrative compulsion was not possible in a UN territory which was striving towards independence.

— The emphasis on erosion control proved to be unwise. The population was not willing to co-operate, and could no longer be forced to do so.

— Attempts to rationalize animal husbandry, and especially to improve methods of grazing, failed entirely.

but, on the other hand:

- — Provision for increased personnel and of funds for the Ministry of Agriculture increased market production by the peasants.
- — Encouragement and support was obtained for cash crops — the appeal to make money was effective.
- — Agricultural extension services which concentrated on simple and profitable innovations achieved better results than administrative compulsion, which was no longer effective anyway.

The agricultural development policy of the late 1950's made use of these lessons.

III. The Late 1950's: A Strategy for Agricultural Development

In the final years of British rule in Tanganyika a coherent agricultural policy, which might be called "a strategy for agricultural development", was worked out. This was hardly a sudden change of mind. The procedure was simply one of "trial and error". Successful approaches were put forward in rather a pragmatic way, while those which failed were dropped.

The outcome of this process of adaptation was an agricultural development policy which relied mainly on:

- — *Improving existing peasant farms via the persistent persuasion of an expanding agricultural extension service.* Thus administrative ordinances as a development tool were replaced by voluntary co-operation with extension officers. It was up to the farmer to decide whether he liked extension advice or not.
- — *Transition to farming under specific schemes, where close supervision was exercised.* Farmers who participated in specific schemes — irrigation, plantation crops, ranching — had to follow certain rules. There was no direct compulsion. Farmers were free to participate or not. Those who took part, had to follow the rules.
- — *Organization of sales through co-operatives and marketing boards.*

This threefold approach gathered momentum in the late fifties. The British expected it to last into independence. They hoped to support it through development help from British and non-British sources. In the following pages the British approach will be described in some length. It is of general value, because it can be considered a useful contribution to the practice and method of introducing "improvements, innovations and new combinations" into tropical farming. Of course the story of what the British wanted to do is — with regard to Tanganyika — largely economic history. Independence brought changes. Nevertheless it can be assumed that a significant part of the British approach is still active in Tanganyika, partly because British personnel are still working there, partly because African personnel schooled by the British have taken over.

1. Extension Work on the Principle of "Persistant Persuasion"

a) Basic Considerations

The new approach to agricultural development is based on the fact that rapid and cheaply accomplished increases in production are necessary, and that the key thereto lies in activating existing peasant farms. The spread of improvements and the introduction of new ideas can undoubtedly multiply production. Essential is not so much capital investment as careful husbandry: early planting, timely weeding, proper tree management, use of veterinary facilities, interest in new crops and varieties etc.

Compulsion is no longer practicable. There is consequently no other way but to appeal to the farmer's self-interest in adopting proven and adapted innovations. Certainly, it is true that farmers like to stick to old habits, yet — and this was stressed by MALCOLM 30 years ago — there is hardly anything in tribal heritage which works against the use of better farming methods. The situation is different from that in India or Madagascar where we can find a definite cultural heritage hostile towards economic innovations. In Tanganyika this may be true for animal husbandry, but hardly for crop production. The problem lies not so much in overcoming cultural obstacles but in mistrust, fear of risk, ignorance and apathy. Persistant persuasion by agricultural extension workers is trying to overcome these barriers. The peasant desires more cash and a gain in social status. Through agricultural extension these desires are used to obtain more production and thus further economic development.

b) Objective: More Cash Crop Production

Priority is given to cash crop production, with additional considerations concerning the subsistence of farm families. Erosion control becomes a secondary item. The important short term aim is for every family to cultivate one main cash crop. In the long run the endeavours aim — as far as soil and climate allow — at establishing mixed farms, equipped with oxen and plough, at cropping in rotations, growing fodder on arable land, using manure etc. Since Tanganyika's inland marked is small, the preference for cash crops is largely the same as that for export crops. The explanation is obvious. The economic development of Tanganyika depends greatly on measures taken by the government. As tax revenues increase, more development services can be provided. For this reason agricultural extension should be self-financing, i.e. should at least be able to cover costs by effecting more market production and thus more tax revenues. Extension work which is mainly concerned with improving the subsistence economy can have only a small and indirect effect on tax revenues. In order to cover the development outlay through increased tax revenues so that more development efforts can be financed, increased division of labour, increased sales, processing, export and import are necessary, i.e. the farmers must produce more cash crops.

The priority given to the production of cash crops is also justified in the light of other considerations. Where no cash crops are produced the farmers have no opportunity to obtain money other than as wage workers in towns and on plantations. Yet towns and plantations are not in a position to absorb the growing labour force. The promotion of cash crop cultivation offers another possibility for earning money and, therefore, indirectly takes some of the pressure off the labour market.

An expansion in the cultivation of cash crops is — paradoxically — an insurance against famine. Protection against famine is not so much a question of extending the cultivation of such crops as the household raises in any case. In good years, surpluses are harvested, and there is no profitable way to make use of them. In dry years, even the cultivation of still larger areas does not provide sufficient food supplies. Having a cash crop means to have another plant, with different requirements as to soil and rain. This helps to reduce the hazards. It may safely be assumed that the more successful third of the farmers save money and bury their savings. In poor years they can then fall back on them.

The promotion of cash crop production is probably in the long run also a better starting point for improving the nutrition of farm families. The idea that the development cycle begins with better nutrition, followed by more work which then leads to increased production should be regarded sceptically. Observations in Tanganyika indicate just the opposite: increased sales lead to great division of labour, to familiarity with other usages, to a gradual change in economic receptivity, to a growing willingness to accept innovations, including those of giving one's self and one's family better food.

c) The Means: Introduction of Popular Innovations

Considering the demand for the fiscal profitableness of agricultural extension on the one hand, and the inability to effect innovations through the use of administrative compulsions on the other, the most suitable method undoubtedly lies in concentrating extension on innovations and improvements which are willingly taken up by the farmers. In an attempt to systematize actual experience in Tanganyika, three groups of innovations may be distinguished:

Innovations which are easy to introduce. In this group are innovations which do not require far-reaching changes in living patterns. They offer a favourable input-output relationship, are technically simple and can be realized within a short period of time. These features are more or less characteristic of such innovations as better seeds, new varieties and new cash crops. The use of proven insecticides also belongs to this group.

Innovations which are difficult to introduce. The second group includes innovations the success of which depends upon their being used in combination with others. This is especially true of measures to increase yields per acre. Efficiency of mineral fertilizers depends upon good husbandry of soil and plants. Systematic livestock feeding requires high quality animals, which are not profitable without the cultivation of fodder crops, the conservation of fodder and veterinary care. An example of the interdependence of various innovations is evidenced by the displacement of the hoe by the ox-plough.

New Systems of Farming. A transition to new systems of farming is even more difficult. The integration of crop farming and animal husbandry, via fodder crops, dairy cows and manure will occur, at best, gradually. Nor can we expect that the initiative of inexperienced peasants is enough to handle newly irrigated land.

Tanganyika is fortunately in a position where it is possible to promote more of the easy-to-introduce innovations. Agricultural extension makes use of this possibility. One of the most important tasks is the introduction of new varieties and new crops. The extension of difficult innovations, i.e. raising the yields per acre by better husbandry, remains largely restricted to some favourable areas. Efforts to introduce mixed farming have been limited to a few experiments. Altogether, this procedure corresponds to an "out-cropping" of those innovations which stimulate the initiative of the farmer. Other innovations and improvements, though they may be technically and economically advantageous, will have to come later.

d) Instruments for the Spread of Innovation

Trained Personnel. The major instrument of such a development policy is trained personnel. The employment of great numbers of untrained instructors can be a rational measure, if carried out within the framework of ordinances and supervision. The new development policy needs extension personnel who can make rational decisions on their own. For this reason the expansion of the staff of the Ministry of Agriculture, and especially of the Extension Division, receives priority in the distribution of government funds. The agricultural education system is being expanded. It is assumed that Africans, who are familiar with the language and customs can, under initial guidance by experienced agriculturists, perform better extension work than Europeans. They also cost less. The aim is to have one officer at the disposal of approximately every 1,000 farmers.

Information. Persistent persuasion in reference to agricultural progress begins at school. Upper primary schools often have a teacher for agriculture and a school farm. Local festivals are held in conjunction with small agricultural shows and contests. Pamphlets and placards on important cash crops are distributed. A weekly agricultural newspaper, Ukulima wa Kisasa (Modern Farming) is published and bought. It is the Swahili paper with almost the highest circulation (about 20,000) in Tanganyika. The radio broadcasts agricultural information. Farmers listen to it in local shops. Extension Officers are equipped with pamphlets, coloured slides and in some cases even with films.

Demonstration. Information is supplemented by demonstrations through experiments, experimental farms and especially through the example of more progressive farmers. Agricultural Officers of various districts set up "Trial Demonstration Farms", i.e. small farms, of the usual size of those in the district, which are farmed with the equipment ordinarily used in the area. Moreover, modern yield-raising techniques (seeds, fertilization, insecticides, good husbandry, etc.) are applied.

Opinions about Demonstration Farms differ. Often Agricultural Officers do not have enough time to take proper care of them. They are run by trainees or assistant field officers, who are not always equal to the task. The prerequisites for demonstration purposes are not always met. In any case a demonstration on such a farm is not nearly as effective as demonstrations on the neighbour's field.

Assistent Field Officers and Instructors who live in the villages are supposed to put into practice the theory which they teach, i.e. to cultivate their own small fields according to the rules of good husbandry. The "baraza", i.e. the village meeting, where the Agricultural Officer used to proclaim ordinances, is replaced by extension work with individuals or with groups of progressive farmers.

Up to the mid-fifties, Agricultural Officers mainly worked with headmen and chiefs. That explains some of the setbacks. Headmen and chiefs carried local administrative authority. Yet the local people regarded many of them neither with respect, nor did they refer to them because they had little or no traditional standing. Headmen and chiefs had been nominated and replaced by different colonial administrations. Frequently they were little else but buffers between a foreign overlord and the indigenous population. Actual authority lies in the hands of individuals who might be called "hidden leaders". Most of them have no apparent power. Sometimes their influence is based on wealth, i.e. on cattle. More important is the respect accorded to members of old families with some traditional standing. Decisive is usually strength of personality.

Extension work with small farmers is dependent on group reaction. Otherwise economic progress is insufficient. In order to get group reaction, the co-operation of the "hidden leaders" is essential. Extension officers are therefore expected to approach the interested and the important ones in the village and to group them together as progressive farmers. In some areas "Progressive Farmers Clubs" are established. Certification from the local agricultural officer is necessary for membership. The grouping together of the progressive ones might be called an attempt to establish rural élites with progressive attitudes.

Progressive farmers should show others that it pays to look for innovations, that much more can be earned in one's own field than from wage labour on estates. The progressive farmers are helped with small loans, close advice and to build up a reputation through high-ranking visitors, newspaper reports etc. Extension work is thus exploiting the amazing difference which exists between good and bad, industrious and lazy, intelligent and stupid farmers.

Autonomous Multiplication. The problem of extension work in a small peasant community lies in the relatively little direct effect on production. One field officer can supervise at the most twenty farmers, which means that for farms averaging four acres, no more than eighty acres can be under his influence. The following example shows the fiscal profitableness of such a situation:

Input: 1 Assistant Field Officer, including equipment and costs of supervision annually = 6,00 shs or 30,000 shs for five years.

Output: It is assumed: In five years the officer is able to increase gross returns per farm from a basis of 700 shs by 120 shs annually, so that in five years the returns equal 1,300 shs. The additional expenses per farm increase from 20 shs in the first year to 100 shs in the fifth year. In these five years then, due to the work of the extension worker, each farm receives an additional income of 1,500 shs, and the twenty farms together an additional income of 30,000 shs.

According to tax figures used in calculations of the World Bank Report, the following additional incomes and taxes are to be considered:

1. Additional farm income	30,000 shs	
12 per cent of this is going to the Treasury via purchase tax		3,600 shs
2. Additional income of commerce and transport amounts to 20 per cent of the additional farm income	6,000 shs	
The Treasury receives 20 per cent thereof out of different taxes		1,200 shs
3. "Primary income" total (1+2)	36,000 shs	
4. "Primary income" stimulates other economic activities and creates "secondary income". The World Bank Report estimates "secondary income" for export products as 1.5 times the primary income, and for non-export products at 0.2 times the primary income. In this case 50 per cent of the additional returns are made up of export products. The secondary income therefore amounts to	30,000 shs	
taxed at 17 per cent		5,200 shs
Total Additional Income	66,600 shs	
Total Additional Taxes		10,000 shs

The outlay of 30,000 shs for extension work is therefore accompanied by additional revenues amounting to not more than 10,000 shs. The fiscal profitableness of extension work with individual farmers is therefore dependent upon the following factors: (1) that the neighbours follow the example of the progressive farmers, (2) that the extension worker, after having helped twenty farmers for five years, takes on new farmers, and (3) that the farmers in the original group of twenty continue on their own to increase production.

If, for example, the extension worker takes on another twenty farmers in the next five years, and achieves the same results, and if the original twenty farmers continue on their own and increase their incomes by 100 shs per year, then in the next five years the extension costs of 30,000 shs will effect tax returns of approx. 20,000 shs. Also there are then forty farmers, instead of twenty, who can demonstrate. A full coverage of the extension costs could be achieved in the second five-year period, if each of the forty farmers could induce his neighbours to increase production to the value of 1,700 shs. The tax revenues would then be sufficient to cover the negative balance of individual counselling.

Such calculations are extremely speculative. They indicate, however, that small farms and low prices are unfavourable conditions for extension work. It is, therefore, necessary to concentrate firstly on export crops, for these have a high multiplier as regards "secondary income" and tax revenues — and secondly on the popularization of simple innovations. Persistent persuasion can only be profitable if innovations once proven and established spread by themselves.

This means that innovations have to be tried out first. For example, the introduction of virus-resistant varieties of manioc which bring higher yields takes place in the following four stages: First there is the work of plant breeding and field experimentation. Then follows cultivation on experimental and demonstration farms, then public experiments on school farms and the farms of progressive farmers. Clearly better varieties can be quickly popularized. An officer in Tanga estimates that 7 per cent of the farmers try the new virus-resistant variety of manioc in the first year following the demonstration. In the second year 14 per cent try it and in 5—6 years 50 per cent. The remaining 50 per cent are not receptive to innovations of any kind.

Activating the People[1]. The autonomous multiplication of locally introduced innovations, which is essential for successful extension work requires the willingness of peasants to accept changes, as well as psychological and social preparation. Information and demonstrations are therefore occasionally supplemented by activating activities. Extension workers and staff members of the "Community Development Department" work together, the latter giving instruction in reading and writing and, for women, instruction in sewing, cooking, home furnishing, and health. They arouse additional wants and needs. What to start with is chosen with care. School attendance is highly regarded. Learning to read and write awakens the wish to send the children to school. At the same time money is needed to buy clothing and to pay the school fees. This means that more income is necessary, and therefore the peasant takes more interest in growing cash crops and is induced to listen to the advice of Agricultural Officers.

---

[1] These efforts are well described by the French words: "Animation rurale". See Goussault, Z.: Report Sur l'Animation Rurale. Tananarive, Madagascar 1960. (Mimeo.)

Farmer's Training Centres. The work of information, demonstration and "animation rurale" in a peasant community can be rather expensive in comparison to the returns. Transportation is costly. Much time is lost on poor roads over long distances. Therefore extension work on the field is supplemented by so-called Farmer Training Centres. Tanganyika is following in this respect the example of Kenya, which already has several such institutes.

30—60 farmers, accompanied by their local extension man, take part in courses lasting from 1 to 3 weeks. Courses are also planned for village teachers, headmen, co-operative society members, youth-groups and party-secretaries. The participants are to be instructed mainly about cash crops which are relevant to them. Teaching is done by resident Agricultural Officers.

Demonstrations take place on a Trial Demonstration Farm. Neighbouring farms, both the good ones and those in poor condition, will be visited. The local extension worker has to follow up the impetus given by the course. Participants should be neighbours where possible, since communal introduction of an innovation spares the individual from having to make an isolated venture. Furthermore, certain innovations (irrigation, dipping facilities against East Coast Fever, cultivation of a crop threatened by wild life, etc.) are only useful if supported by the community. Also, the conformist behaviour of a group of farmers establishes a stronger nucleus for agricultural development. A fee of a few shillings per participant is designed to keep the centre from becoming crowded by disinterested persons. The fee is also intended to show that the services offered are worth paying for. On the other hand, the display of informative material and the demonstrations, with films for example, are designed to make participation attractive. The success of these centres will greatly depend upon whether they can be made popular.

Providing Loans. Loans constitute another important instrument. It costs money to put new ideas into effect. Extension works on the principle that no "gifts" should be made. The free distribution of seeds, planting materials, insecticides, tools, etc. is frequently misunderstood. It leads to dependency and stifles individual initiative. The guiding principle is that farmers have to pay for anything they receive, except for extension itself. In many cases, however, the farmers do not have the money, or — and this is probably more important — they are not willing to risk their own savings. Small loans are therefore indispensable. Allowance has to be made for difficulties arising in connection with the use of loans and the fact that re-payments and the checking of applications is a costly and complicated matter.

A committee of five respected and successful farmers decides whether a loan should be granted. The committee is appointed by the local Administrative Officer and advised by a Loan Executive Officer. Loans from 100 to 4,000 shs can be granted by this committee, which has a special fund for the purpose. Loans may be given only for productive investments. Upon submission of the application a fee must be paid, which is designed to prevent the committee from being flooded with unreasonable applications. The applications are then examined by the Loan Officer and a decision is made by the committee. The committees have in general performed their work well. They do not hesitate to turn down the majority of the applications. So far, securities in land are required only in Chaggaland. In general, such property as bicycles, sewing-machines, livestock, etc. is acceptable. Security may also be offered by presenting two guarantors, or by depositing 10 per cent of the credited sum with a Credit Agency. Character and qualifications of the applicant, as judged by his reputation in the community, are decisive elements in obtaining small loans. Interest charged for investments amount to 7.5 per cent annually. Interest charge on crop loans is 8.5 per cent. Most of the applications are made for crop loans. Loan requests for investment in machinery, oxen, ploughs, and the establishment of permanent crops are much less frequent.

Local Authorities. Local Authorities also play a different role under the new agrarian concept. They no longer need to issue agricultural ordinances and to punish offenders. They are no longer the employers of Agricultural Officers and instructors, as they were under the Sukumaland Development Scheme. The Agricultural Department has full responsibility for extension work. The Local Authorities remain, however, responsible for local investment in wells, dams, and small irrigation works. As far as tree nurseries, spraying of crops, etc. are concerned, agricultural personnel carry out these services, but Local Authorities remain financially responsible.

e) Operational Procedure: Concentration of Effort

The main principle of extension work is the concentration of effort. The danger of "the seeping away of initial success" is greater in Tanganyika than in many other underdeveloped countries, because of the widely scattered population and the small number of extension workers. Concentrated efforts must achieve cumulative development in a given area, overcome "threshhold values", break down "barriers", and eliminate prejudices and mistrust through persistent repetition. Efforts must go forward from "bridgeheads" of agricultural progress, because holding-on, keeping-up, and overcoming unexpected difficulties is usually even more difficult than introducing a given innovation. Therefore the personnel and money in extension work is concentrated on:

— densely populated areas, where soil and climatic conditions are favourable and markets are assured. Land shortage is usually the most effective forerunner of agricultural progress. Where additional land is no longer available, the only alternative is to raise yields per acre. The willingness to co-operate with the extension worker develops at the same time. Rounce wrote therefore: "A situation of overpopulated villages and impoverished soils must be passed, before better husbandry can spread."

— receptive farmers, villages or tribes. In this connection there are many variations, even under similar climatic and soil conditions. Peasants are usually more receptive to changes than herdsmen. Some of the cattle-herding tribes cling so tenaciously to tradition that extension work seems to be a hopeless measure. Extension work is concentrated on a few interested cash-crop-minded tribes.

— specific projects. Local Agricultural Officers are responsible in all matters of agricultural development. They have to perform numerous and various duties. But, as far as possible, their work and that of their staff is concentrated on one cash-crop in a few villages.

The concentration of agricultural development work is a necessary consequence of the conditions of production in Tanganyika. A staff which is widely scattered throughout the land, largely inexperienced, and in need of constant encouragement, can hardly be supervised and instructed. High transport costs and time lost in travelling long distances must be avoided. Concentration of personnel and money helps to realize those objectives which, in terms of outlay and returns, are most promising.

This, of course, leads to the fact that agricultural development efforts do not help all farmers in the same manner. Experience shows that aid measures are more effective if carried out in more developed areas. The economic usefulness of extension depends upon the condition that help is given to the person who helps himself, and such a person is generally already in a relatively good position. However, Tanganyika is only at the beginning of its agricultural development, and the existing small islands of agricultural progress must be strengthened and encouraged. In the course of time they may spread out and finally cover the whole country. In the beginning the outcropping of favourable locations must be considered as unavoidable.

f) Examples of Successful Extension Work

The basic characteristics of extension work — as developed by the Agricultural Department shortly before Tanganyika attained independence — are illustrated by the following examples. In September 1956 the African members of the Legislative Council presented a tentative 5-year plan for the development of agriculture in Tanganyika. The aim was to finance additional social services by increasing tax revenues. The spur came from the so-called Increased Productivity Schemes. Following the principle of concentration, 140 specific projects limited to small areas were considered. Work was begun on 98 of these projects, and extended on 85 of them (including forestry and fisheries). In 1960 seven additional projects were begun. Most of these projects concentrated on the introduction or improvement of relatively profitable cash crops: coffee, tobacco, pyrethrum, cocoa, seedbeans, tea, etc. Some of the projects failed, others came to a standstill. All in all, the undertaking proved highly successful.

Example 1: Expansion of Coffee Production

There were numerous projects concerning the expansion of coffee cultivation. In every case the projects were carried out in areas other than the traditional coffee-growing area of the Wachagga. The procedure was: An Agricultural Officer is stationed in a climatically suitable area. He is supported by assistant field officers and/or instructors. Nurseries and fields are established. Extension workers explain to the farmers that it is profitable to plant more coffee, to buy young plants of good quality from nurseries (where necessary, on credit), and that it is worthwhile to husband plants carefully [1]. The results of this work can be seen in the figures in Table 23. TACTA incorporates various small groups of coffee planters who work in different parts of Tanganyika, and who started or increased output as a result of the Increased Productivity Schemes. Since the introduction of these schemes in 1956/57 coffee production has almost doubled in spite of falling prices.

Table 23. *Coffee Production Figures of the Members of the Tanganyika Co-operative Trading Agency (TACTA)* [1]
(tons of raw coffee)

Year	Production in tons	Price £/ton
1952/53	819	464
1953/54	1127	612
1954/55	1252	415
1955/56	2067	468
1956/57	2081	492
1957/58	2851	368
1958/59	3156	310
1959/60	3792	315
1960/61	4096	286
1961/62	3949	290

[1] Production areas: Rungwe, Matengo, Meru, Koimeri, Mbozi, Usambara, South Pare, Kindoroko, N. C. U., Lupembe, Ubena et al.

Source: Taken from: TACTA: Managers Report for the Year ended 30th June 1962; Mimeo.

Example 2: Introduction of Pyrethrum [2]

Up to 1956 Tanganyika's pyrethrum came almost entirely from the estates. Demonstration on estate fields was not enough to induce the peasants to produce this profitable crop. Agricultural extension was necessary. It yielded rapid and amazing results. In 1962 the number of pyrethrum growers already amounted to 60—80,000. Expansion continues. In 1963/64 sales of 3,500 tons dried flowers are assured. It is not unlikely that more will be offered.

Pyrethrum cultivation by peasants started with the search for a cash-crop in those mountain areas which are unsuitable for coffee. Mountain farmers usually cultivate maize with returns of hardly more than 150 shs/acre. Pyrethrum, however, with a price of 2.50 shs/lb dried flower and yielding 250 lbs dried flower/acre gives returns of 625 shs/acre. There is the example of Kenya's pyrethrum pro-

[1] Since the signing of the Coffee Convention in 1962 the work is limited to improving the quality of existing coffee trees.

[2] Pyrethrum, which is obtained from dried flowers, is the most effective natural insecticide. World production in 1958 was 11,000 tons, 5,000 of which were produced in East Africa — mostly in Kenya. The USA bought 63 per cent of world production. In 1962 there were marketing difficulties. Attempts to control production would strengthen Kenya's position, but would stifle Tanganyika's development, and are consequently rejected in Tanganyika. (As to pyrethrum cultivation see: Tuckett, J. R. Pyrethrum, Bulletin N. 5, 1961.)

duction by the peasants. It is the ideal peasant crop. No investment is required except for clearing the land and buying the splittings. Pyrethrum, however, needs much manual labour: weeding, picking and drying.

The introduction of pyrethrum in Njombe District is a suitable example to demonstrate the different steps taken in order to introduce a new crop:

— First comes scientific research: examination of soil, climate and varieties. Initiative for research goes back to questions asked by the local chief, who wanted a cash crop for his far-off area.
— Barazas are held. The chief explains the new crop. He refers to some European estates in the area which grow pyrethrum. He stresses the point that they make money with pyrethrum.
— Initial interest is negligible. Ten persons decide to plant one quarter of an acre. Splittings are supplied free of charge. Planting and husbandry is closely supervised by extension personnel. They need an initial success.
— The first harvest is bought on the spot and later on destroyed by agricultural personnel. The small volume of production does not justify marketing expenses.
— One of the first farmers to try cultivating pyrethrum is a former foreman on a European pyrethrum growing estate. It is noteworthy that initially interest does not come from independent peasants, but from farmers who may be termed small agricultural entrepreneurs because they employ seasonal wage labour in order to grow some pyrethrum.
— Interest is growing. The Agricultural Department establishes nurseries. They are financed through the Local Authorities and run by the local Agricultural Officer.
— Splittings are sold to potential growers, either for cash or on credit. In some exceptional cases they are distributed free of charge.
— Interested farmers receive a licence to grow one acre of pyrethrum and not more. The licence is liable to be cancelled in case of bad husbandry. Licensing is considered an instrument to control husbandry as well as supply.
— Extension work in the area is concentrated on pyrethrum. Introduction of this new crop is coupled with a number of other improvements — terracing, regular weeding, and the application of mineral fertilizers. Experience shows that farmers who do not know the new crop are willing to grow it according to the rules laid down by extension workers, something they would not even think of doing in the case of old established crops. Thus the new crop is used as a lever for innovations which are of general importance. With increasing numbers of pyrethrum growers, co-operatives are established. It is intended — but this is usually too much for them — that they run their own nurseries. Co-operatives, however, carry the licence for a certain pyrethrum acreage. It is up to them to distribute this acreage among the interested ones.

In some cases pyrethrum growing is made attractive by providing forest land. Growers receive one acre of rough bush belonging to the forest reserve. They can keep it for five years. It is cleared, and planted with pyrethrum. In the second year of clearance, pines are planted between the pyrethrum rows, thus establishing a mixed cultivation of pines and pyrethrum, with the pines profiting from pyrethrum-weeding. After five years the pines take over, and the pyrethrum growers receive new plots.

The problem with pyrethrum is that markets are limited to definite amounts. Yet it need not be feared that development stops as soon as these limits are reached, that the introduction of pyrethrum in a certain locality will end with "stagnation on a somewhat higher level". Pyrethrum sales frequently spark off a peasant community. It can be observed that interest in other cash crops intensifies as soon as there is a rumour of pyrethrum licences being limited.

Example 3: Higher Cotton Yields per Acre in Kilosa District

In some areas of the Eastern Region soil and climate are especially good for cotton. Non-irrigated cotton yields up to 2,000 lbs/acre if

land is carefully prepared,
planting is done early (February) and on ridges,
intervals between rows are correct,
thinning is properly carried out,
weeding is done early and often, and
insecticides are properly applied.

To get this done a scheme was initiated, which was called "Insect control for cotton in the Kilosa and Morogora districts". The scheme started in two villages, Ulaya and Ukutu. Later, a village in Bagamoyo was included. The scheme began in 1960, making use of the preparatory work of earlier Agricultural Officers. Farmers reacted in the first years. Between 1934—1960 seed cotton production at Ulaya fluctuated around 60 tons. In 1962 production reached 300 tons and yields per acre 1,000—2,500 lbs of cotton. The larger crop is almost entirely due to higher yields per acre. The procedure in Ulaya is as follows:

Preparation: Thanks to thirty years of work done by the neighbouring Ilonga research station, well proven recommendations can be made concerning the cultivation of cotton. There are already a few receptive farmers in the village. The neighbour's example is at hand. Six assistant field officers and a Community Development assistant, as well as a health assistant, are chosen for the work. The extension workers receive refresher courses in cotton cultivation at the research station. They are trained in methods of community development at Tengeru, the agricultural school near Arusha. The important personalities of the village, including the TANU secretary, undertake in addition inspection and information trips to Tengeru.

Extension Work: Extension workers concentrate on one crop, that is, on cotton. They work with groups, using projectors, posters and films. They arrange field inspections, and finally visit the farmers who are especially interested. Due to an extension density of 200 farmers for each officer, the interested parties, about 10 to 20 out of a hundred, can be given personal advice. Agricultural extension is supplemented by general instruction in reading and writing. "The Ten Commandments of Cotton Cultivation" is used as a textbook. In addition to special scheme workers, the Agricultural Officer of the district devotes most of his time to cotton in Ulaya. The Community Development Assistant and the Health Inspector provide information for improved living standards through health control, useful nutrition, tin roofs, etc.

Self-Help: The most important aspect of self-help is the willingness to form small producer's co-operatives. They provide the nucleus around which the progressive elements assemble. The co-operative societies are distributors of insecticides and of sprays, which are given on a credit basis. A further form of self-help is the construction of three community centers. The government grants £ 100 for each building. The village population executes the construction work. The attainment of political independence has contributed a great deal to wards people being prepared to accept economic changes.

Outside Help: The government covers the extension costs and contributes £ 300 for the construction of the community centers. Credit is given by the "Lint & Seed Marketing Board" at a rate of interest of 7.5 per cent. The co-operative society acts as guarantor, and transmits the credit only to those farmers whose fields are so well cultivated that the use of insecticides is probably worthwile, in the opinion of the extension workers. Repayment is effected without difficulty when settling for the cotton sold.

Fiscal Profitableness: The economic success of such a project cannot be judged on the results of the first years. A rough calculation of the balance so far shows the following picture.

Input:	8 extension workers each at 4,000 shs annually		32,000 shs
	1/2 Agricultural Officer "24,000 shs"		12,000 shs
	Transport and other costs of extension		10,000 shs
			54,000 shs
	or 162,000 shs in the first three years.		
Output:	1960 preparation.		
	1961 Crop increases in comparison to the previous year		106 tons
	1962 Crop increases in comparison to the previous year		230 tons
			336 tons
	336 tons of cotton at 55 cts./lb		413,950 shs
	less subsidy charges of 7.5 cts./lb		56,450 shs
	less insecticides		50,000 shs
	additional net income		307,500 shs
1.	Additional net income	307,500 shs	
	Govt. receipts in taxes 12 per cent		36,900 shs
2.	Additional income of transport, commerce, and processing industr. 20 per cent of item 1.	61,500 shs	
	of this the Treasury receives 20 per cent		12,300 shs
3.	"Primary income" total (1.+2.)	369,000 shs	
4.	According to the report of the World Bank the "secondary income" for exported products can be calculated at 1.5 times the "primary income", that is	553,500 shs	
	Govt. receipts in taxes 17 per cent		94,095 shs
	Total additional income	922,500 shs	
	Total additional taxes		143,295 shs

Fisal costs of 162,000 shs have brought additional tax revenues of 143,295 shs. By this, even in the first years, the scheme was almost paying its way. The favourable result is due to the fact that cotton is an export crop with a high multiplier. The high fiscal profitableness is also due to the combination of individual and group extension work. It obviously gives better results than does individual counselling alone. The calculation presented here may be in many ways too optimistic. Expenditure and output or multiplier would have to be much less favourable in order to show that this kind of extension work is fiscally unprofitable.

One must remember, however, that Ulaya achieved the best results. In Ukutu the scheme is not progressing in as striking a manner as in Ulaya. In Bagamoyo an initial failure was registered. The cause should be examined. In 1961 there was a poor crop in some parts of Tanganyika. Maize from the USA was shipped free of charge to Dar-es-Salaam to alleviate famine. Other areas felt they also had a right to some of this maize. The village of Bagamoyo therefore requested an allotment. This request was denied. Thereupon the local co-operative refused to co-operate.

Example 4: Integration of Animal Husbandry and Coconut Growing in Tanga.

The coastal people of Tanga region have owned and used cattle and coconut palms as long as man can remember. Yet they never combined these two economic activities. Palms are owned by individuals whose trees are arbitrarily located among the palms of others. The palm owners have rights only of the soil which is in the immediate vicinity of the trees. It is used in irregular cycles, usually to grow manioc, and is then abandoned to grass and bush. Fire frequently burns grass and bush, damaging older palms, killing younger ones and often destroying manioc fields. Coconuts are picked too early because the owners fear theft. The quality of copra is consequently low. Usually the cattle are not owned by the men who keep them. These persons in turn, leave the cattle in the care of young herdboys all day. Grazing starts as late as about 10 a.m.; the animals are driven 6 or 10 miles to watering places during the heat of the day. As a result, milk per cow hardly reaches 100 gallons annually. As a rule cows are milked by one of the numerous milk retailers, who travel on bicycles from village to village, milk a few cows and make deliveries to customers in town. Great losses are caused by disease, especially by the tsetse-fly and East Coast fever. It is assumed that only about 25 per cent of the new-born calves survive.

The experiences of the Tanga Dairy, an experimental station run by the Veterinary Department, show that coconut palms and cattle need not be as unproductive as they are at present. The integration of cattle and coconut growing is to the advantage of both activities. Bush-clearance and a higher grazing density are favourable to the growth of valuable grass. Fire no longer damages palms. Yields per tree rise from 15 to 28 nuts annually. Fencing palm tree areas allows nuts to ripen and fall naturally. Night grazing of the cattle becomes possible. Improved pastures lead to higher milk yields. Zebu cows at the Tanga Dairy produce 500—600 gallons of milk annually. The veterinary care reduces loss through disease from 7—8 per cent to 2—3 per cent of the herd annually. Due to these achievements, and to the high milk price of 5 shs/per gallon, the Tanga Dairy was able to make a profit a few years after it started.

The success of Tanga Dairy led to the question as to whether the technical progress achieved there could be transmitted to peasants. The city of Tanga and the numerous workers on the neighbouring sisal plantations provide a big market for milk, meat, and coconuts. An attempt to answer this question has been made by the Cattle-Coconut Co-operatives which have been organized through the efforts of two Agricultural Officers, A. H. B. Childs and C. G. Groom, and their assistants.

The procedure is as follows: a group of interested farmers is asked:

— to divide their land into two parts, coconutpalms combined with grazing on one side and arable land — with or without palms — on the other.
— to graze cattle communally and to milk cows in a communal milk-shed.
— to clear bush in the grazing area.
— to follow veterinary rules and to pay for medicaments.
— to castrate bulls.

The participants receive:

— free veterinary supervision.
— a grant, which need not to be repaid, for a water pump and fencing.
— a high quality bull for some time. The bull remains the property of Tanga Dairy.
— easier access to small loans.

Cattle and palms and their returns remain private property. Outside the grazing area, the peasants continue to grow crops as they are accustomed to. They are, therefore, independent farmers who submit voluntarily to some kind of land consolidation and communal land use. The first scheme began in 1959. In early 1963 the number of cattle-coconut schemes already amounted to 43.

The first and most successful co-operative of this kind has 75 members and covers 750 acres and 250 head of cattle. In the initial stages apathy was apparent. The visit to the Tanga Dairy did not create much of a change of mind. The free chemical clearance of some bush left little impression. The construction of a watering place, which made it easier for the women to obtain water, established some confidence. The veterinary supervision of the village cattle added to the interest. Only 3 per cent of the cattle died each year in comparison to 25 per cent previously. Political independence gave a strong impetus to co-operative efforts. The recognition given to the initial achievements by numerous visitors was very helpful. The inhabitants began to think of the scheme as their own achievement. In the initial stages the apathetic farmers had laconically replied, "We want to live, not to work". As time passed the scheme became an object of self-help. Additional bush was cleared manually by those taking part. They constructed another watering place and drinking facilities for the cattle by themselves. The Agricultural Department supplied only the materials.

Finally success became obvious to all. Coconut palms and crops are no longer endangered by fire. New palms can be planted without the risk of their being destroyed. The cattle is put to pasture at 7 a.m. instead of 10 a.m. Fodder is better and more plentiful and larger herds can be kept. As dips are available, participants can obtain small loans in order to purchase more cows. Higher quantities of milk are obtained and can be sold for 3 shs per gallon. Childs and Groom record the usual family income for this area at 540 shs annually. After 3 years of participation on the cattle-coconut project the family income has risen to 1,500 shs.

The economic balance of one scheme can be roughly estimated:

Input:	Grants for pumps, fences, etc.	2,720 shs	
	1 Field Assistant Officer		4,000 shs
	Material costs for 1 Officer		1,000 shs
	1/5 Agricultural Officer		5,000 shs
	Initial Investment	2,720 shs	
	Current annual costs		10,000 shs
	Total costs in the first 3 years		32,720 shs
Output:	1 year of preparation		
	2nd year — increased production per participant in comparison to the previous year .		500 shs
	3rd year — increased production per participant in comparison to the first year . . .		1,000 shs
	Total		1,500 shs
	or, for 75 participants	112,500 shs	
	less material costs of 20 per cent	22,500 shs	
		90,000 shs	
1.	Additional farmer income	90,000 shs	
	Gvt. tax receipts 12 per cent		10,800 shs

2. Additional income from transport, 20 per cent of item 1.	18,000 shs	
Gvt. receipts in taxes 20 per cent		3,600 shs
3. Total "primary income" (1.+2.)	108,000 shs	
4. It is assumed that 25 per cent of the "primary income" is provided from export crops, for which a multiplier of 1.5 is used, and that non-export crops account for 75 per cent, for which a multiplier of 0.2 is used. The "secondary income" therefore amounts to	56,700 shs	
from which taxes of 17 per cent are paid		9,639 shs
Total Additional Income	164,700 shs	
Total Additional Taxes		24,039 shs

Therefore fiscal costs amounting to 32,720 shs have produced tax revenues of 24,039 shs. In view of the fact that these figures represent the first 3 years of the project, the results are very good, especially since the integration of animal husbandry and coconut growing has long-range effects on coconut yields and grazing quality. These calculations indicate that the cattle-coconut approach had a successful start. The next few years will show whether a lasting change has been achieved. Cattle-coconut schemes require a high degree of willingness to co-operate.

Possible sources of tension can be seen in the fact that bush clearance is done by all, while the profit from this largely goes to those who own cattle. Property in cattle, however, is concentrated in the hands of some wealthier individuals. It may be asked how long those who do not own cattle are willing to participate in communal clearance work. On the other hand the endeavours of the politicians to run dairy cattle and coconutpalms on an entirely communal basis might destroy individual responsibility and the essential relationship between individual effort and individual return. As to milk cows, individual care can be considered indispensable. Another scheme-destroying tendency is the fact that the participants like to employ outside labour for milking and herding. Thus the participant in the co-operative is placing himself in the position of a man who enjoys in leisure the economic rent created by institutional progress.

Whether or not cattle-coconut schemes will last and flourish, will largely depend on the government's policy. They have a fair chance, if gifted extension workers are provided, political experiments are avoided, and unruly individuals are excluded. In any case, the Agricultural Department has an effective "lever". If participants do not carry out the extension workers' advice, then — as has already happened in one instance — pump and fence are removed. Women would have to walk long distances to a well to get their water.

Example 5: Popularization of Ox-drawn Equipment

The attempts of Local Authorities to introduce tractor cultivation by installing tractor stations in the early 1950's failed. Since that time the Agricultural Department has been trying to popularize ox-ploughs. The reasons are obvious: oxen (and donkeys) are already available. Oxen are used in Uganda, Kenya and parts of Tanganyika. The costs are small. The farmer who already owns oxen only needs to purchase a plough. Where marginal land is cultivated, and where gross yields per acre bring returns of only 100—200 shs, the only choice is between ox-plough and hoe.

The returns simply do not justify the cost of a tractor. If farming is done with the hoe a maximum of 4—6 acres can be cultivated, limiting the family income to 400, or at best 1,200 shs. With the use of oxen the farmer can cultivate 10—15 acres, and earn correspondingly. With ox-carts, manure can be transported from the "bomas" to the fields. Pack transport of manure can also be effected with donkeys — a common practice in Greece. The Agricultural Department promotes the use of oxen and donkeys by granting loans for the purchase of ploughs, oxen and carts. The Department runs small stations — of which there were five in 1960 — where instructions in the use of oxen, ploughs and carts are given [1].

In some areas the campaign to popularize ox-ploughs seems to be a full success. The number of ploughs in Tanganyika in 1962 has been estimated at 60,000. Most of them have been put to use in Sukumaland. The use of ox-teams has clearly taken hold here. There is also a small but definite increase in the number of ox-carts.

2. Production under "Close Supervision"

Efforts to improve existing African farms by "persistent persuasion" have definite limits. They must confine themselves to the "outcropping" of popular innovations. It would be useless for extension work to attempt innovations which exceed the farmer's capacity. Popular innovations correspond largely with the expansion of cash crop cultivation. Within the existing system of land use this means increased "soil mining". Erosion dangers are usually emotionalized. They should not be overvalued in a country like Tanganyika. Soil mining can hardly be avoided if existing farms are induced to increase production. On the other hand, loss of fertility should not be overlooked. Its presence indicates the necessity to modify farming in specific cases, i.e. developing rationally organized farms on which soil fertility is preserved, crop growing and animal husbandry are integrated, and modern methods of production are used. Therefore the British conception of production development in Tanganyika contains two elements: first, the increase of production on existing peasant farms by popularizing simple innovations and secondly, the establishment of an initially small, but gradually expanding modern sector.

A "leap" into modern farming technology cannot be expected from peasants who are still bound by tradition. Improvements will probably take place gradually. Areas with land shortages will be the forerunners. In the course of time more use will be made of manure, the cultivation of fodder crops and the integration of crop farming and animal husbandry. But this is likely to be a slow process. Extension advice is not sufficient to effect a rapid transformation in farming. In view of the manifold and incalculable interdependence which goes with the introduction of new, related systems of land use, the risks which the individual would have to face are overwhelming.

---

[1] Some of the ploughs which the Prime Minister of Nordrhein-Westfalen presented to President Nyerere during his visit to Germany are also used here for instruction purposes.

The *transformation* of farming in specific cases, as a supplement to *improving* large numbers of existing farms, calls for "production under close supervision". In this connection the British Administration tried to create situations in which the farmers are encouraged and induced to conform to certain rules. The utilization of additional potential is one of the basic ideas for making this system effective. For example, the agricultural administration can restrict settlement on new land, or participation in irrigation schemes, to those applicants who declare their willingness to conform to the regulations laid down. There is no compulsion. The individual is free to make his own decision concerning participation. The discussion is still going on as to which forms of supervised production under changing conditions are suitable. The experience gained in this second important approach by British agricultural development policy is noteworthy and some exemples are given below.

a) Settlements on Non-Irrigated Land

α) Unplanned Settlements Subject to Certain Rules

The simplest form of supervised production is evident in unplanned settlements on new land. A specific area suitable for farming is opened to settlement. The agricultural administration drew up, for example, the following rules: —

— Only a restricted number of families may establish themselves in the area.
— The number of livestock is limited.
— Farming is only allowed in parts of the area.
— Slopes are to be reserved for grassland and forests as protection against erosion.

Unplanned settlements are cheap. Investment is limited to an infrastructure of roads and water. Individual farms are not planned. The peasants divide the land among themselves and farm it in the usual manner. This form of settlement, which took place — for example — in Geita District and in the Ismani Valley near Iringa, can be highly successful economically. Within a few years production increases significantly. Efforts to transform, however, fail. Experience shows that the rules cannot be implemented. Population density and the number of live-stock cannot be controlled. After a few years, farming is the same as in other areas where no such efforts have been made.

β) Planned Settlements on New Land

In other cases settlement on new land is more carefully planned. The planning consists normally of (1) careful mapping, (2) testing of soil and climate, (3) selection of a clearly marked and circumscribed scheme area, which the Local Authorities place outside of local law and custom, (4) construction of feeder-roads, watering places for man and cattle, etc., (5) the construction of contours on the prospective farm land, so that the settlers are given a framework for erosion control. Buildings are usually not constructed; the settlers erect their own huts; they also bring cattle and equipment; in some cases loans are given for the purchase of ploughs and oxen; a special settlement officer is made available. Such settlements are usually started because — somewhere in the district — there is a relative shortage of land. The Local Authorities are pressed by the peasants, who desire more land. This pressure is passed on to the agricultural administration, which in turn looks for suitable areas and combines the settlement of such areas with ambitious objectives. Farmers are expected to develop their own medium-sized commercial holdings, with homes on the land, based on the integration of crop farming and animal husbandry. Sometimes the land allocated per farm amounts to 100 acres. In some cases hedge sisal along the borders indicates that grazing within these borders is a matter of individual responsibility. In other cases co-operative responsibility for the use, husbandry and maintenance of the grassland is hoped for. Final responsibility for the settlements remains in the hands of the Local Authorties. Participation of the local agricultural administration is limited to planning, organizing, and counselling.

More services are made available on planned settlements than on the unplanned ones and, accordingly, more rules are issued, for example: —

- The settlers must live on the plots which have been allocated to them; livestock may not exceed a certain density, for example 7 acres per head of cattle;
- the settlers may decide on the crops to be raised. They must, however, carry out all of the measures for erosion control which have been laid down by the agricultural administration. Trees may not be cut down without permission, and grass may not be burned;
- the settler must provide for permanent marking of his land by planting sisal hedges;
- the health control for livestock must be carried out according to the recommendations of the veterinarian;
- a committee of leading representatives of the Local Authorities and the settlers is responsible for the implementation of these rules, with powers to expel anyone who does not conform.

So far, experience is discouraging: costs are high, achievements small, and the difficulties great. The services of expensive specialists are needed. A scheme with individual grazing requires — if it is to be successful — the construction of watering places on each holding. The costs, therefore, are prohibitive in relation to the returns. The farmers are not willing to pay for water. They consider it a gift of nature, for which no payment is necessary. Returns from farming are usually mediocre. If land close to heavily populated areas is kept fallow, the reasons are almost always good: low fertility, danger of floods, seasonal flooding, damage due to wild animals, tsetse-fly, etc.

Staff members of various departments participate in the scheme, agriculturalists, veterinarians, engineers, and administrative officials, but no one is really responsible. The schemes are too small and too marginal to justify the employment of a permanent manager whose training and status would ensure a solution for the numerous technical problems and human difficulties. The responsible institutions — the Local Authorities — are rarely in a position to implement the rules which have been adopted. In addition, they are not willing to give the necessary authority and support to the Agricultural Administration. As a rule such settlements do not accomplish much more than the unplanned ones. The additional outlay for preparatory work is unprofitable. Fortunately, only a few of such schemes were begun in the 1950's.

γ) Settlements of the Tanganyika Agricultural Corporation [1]

The projects of the Tanganyika Agricultural Corporation are more significant than the above-mentioned types of settlement. The groundnut scheme left extensive areas of cleared land in its wake. They were to be managed on a commercial footing. The rational solution was the establishment of large farms and an examination of which types of large farms would be most suitable under the existing conditions. The result: state-owned farms of 500—1,000 acres cover a total of 12,000 acres in Nachingwea. In Kongwa the TAC manages a ranch of approx. 90,000 acres. In Urambo most of the land is leased to tenants who operate large farms.

[1] The Tanganyika Agricultural Corporation farms or leases areas in Nachingwea, Urambo and Kongwa which were cleared under the groundnut scheme. It has taken over management of the Mbarali irrigation project, the tobacco farm Lupatingatinga, Ruvu Ranch, and the Sonjo Settlement. Areas managed in 1960/61:

Nachingwea,		12,915 acres cropped
Urambo,	approx.	10,000 acres cropped
Kongwa, settlement		960 acres cropped
Kongwa, settlement	approx.	30,000 acres grassland
Kongwa, Ranch	approx.	90,000 acres grassland
Mbarali, irrigation scheme (1963)		500 acres cropped (aim: 5,000)
Ruvu Ranch,	approx.	78,000 acres grassland
Lupatingatinga, tobacco cultivated on 105 (area 5,000) acres		
Mkata Ranch,	approx.	220,000 acres grassland
Sonjo and Ichonde settlement:		2,000 acres, 50 per cent cleared in 1963

In addition, "African Tenant Schemes" with long-term objectives were started in 1952—1954 in each of the three localities. The report of the Overseas Food Corporation, the predecessor of the Tanganyika Agricultural Corporation, states:

"One of the most stabilizing influences in an African community under the present economic and political pressure being exerted within and without, is a healthy, prosperous, yeoman farmer class, firmly established on the land, appreciative of its fruits, jealous of its inherent wealth and dedicated to maintaining the family on it" (Reports and Accounts, 1954—55, p. 159).

In order to realize these objectives, settlers are established under close supervision. The use of supervised tractors is to do away with the bottleneck of hoe cultivation and is meant to develop the work potential of the peasant families. When working on a well-planned farm, the settler is expected gradually to develop his ability and interest in managing an independent commercial holding himself.

Crop Farming on marginal Land in Nachingwea (Southern Region). In Nachingwea selected settlers are given holdings between 10 and 50 acres. They receive housing, loans for initial necessities, a water supply, schools, a dispensary and the chance to improve their social standing. The scheme management requires a crop rotation. Three years cropping is followed by a fallow period of three years. The management provides for ploughing, ridging, seeds, mineral fertilizers and insecticides. The farming family does the planting, weeding, and harvesting. The cash crop has to be delivered to the management. Payment is made according to previously established fixed prices. The following costs are deducted from the crop returns: (1) the costs of preparing the fields, plus other services, (2) outlay for seeds and fertilizers, (3) 55 shs per acre for general costs of management, and (4) any credit which has been given for the purchase of consumer commodities.

This set-up does not work in Nachingwea. Settlers are not attracted to the scheme. Some interest in the early stages was due to exceptionally good harvests. In 1955—56, 126 settlers were recorded. They farmed 2,506 acres and earned an average family income of 1,146 shs. In 1960—61 there were only 19 settlers. The fluctuation is large. Approximately one half of the new settlers arriving each year leave the project (Table 24). The settlement is not regarded as a home. Participation is looked upon in much the same way as employment on any other agricultural estate. Settlers arrive, stay a few years, save some money, and return to their native villages. The incomes earned are regarded as too low in relation to the necessary adaptation. There is no cash crop which brings high returns per acre. Beans, peas, maize, sorghum, soya beans, groundnuts, sesame and sunflower seeds are cultivated. The gross returns per acre fluctuate between 100 and 250 shs, in spite of European direction, good seeds, and mineral fertilizer. 55 shs of the low gross returns per acre must be paid for the general administrative costs and approx. 100 shs for other expenses.

It is true the settlers average over the years an income of 750 shs. This is twice or three times the income of peasants in the area. In hard times the settlers are supported by the management. They have, however, to sacrifice part of their independence. They must conform to the rules and controls. There is apparently no end to his being pushed around. For almost all of the settlers the prospect of becoming an average farmer lies outside their experience. There is no "pressure situation" in the Southern Region which induces peasants to co-operate.

Land of marginal quality is abundantly available. The needs for consumer goods is even smaller in this isolated area than in other parts of Tanganyika. Participants can always return to the primitive, but not unbearable way of life of the ordinary peasants. They seem to prefer this. Another reason for the failure may be that the relationship between the settlers and the management never developed into one of trust and confidence. The more favourable figures of 1961—62 indicate that perhaps there has been a change recently.

Table 24. *Results of Nachingwea Settlement Scheme*

Year	Number of holdings	Acreage cropped		Total value of crops per holding[2,6] in shs	Expen- diture[3,6] in shs	Holding profit[4,6] in shs	Number of tenants who left scheme at year end[5]
		by tenants[1]	per holding[1]				
1952—53	28	333	11.3	1488	457	1031	16
1953—54	54	886	16.4	3038	1594	1444	14
1954—55	102	1908	18.7	2307	1574	733	49
1955—56	121	2506	21.2	3189	2043	1146	60
1956—57	99	2338	23.6	2987	2452	535	46
1957—58	88	1367	15.5	1904	2088	184	44
1958—59	79	1220	15.4	1528	2104	576	56
1959—60	23	341	14.8	2694	1859	835	13
1960—61	19	261	13.7	2691	1493	1198	1
1961—62	80	948	11.8	2313	1500	813	9

[1] Excluding fallows.

[2] Including value of consumption of foodstuff by farmer families, excluding possible earnings from other sources.

[3] For services and goods provided by the TAC.

[4] Excluding illegal sales.

[5] The discrepancy between the number of holdings and of tenants leaving the scheme at the end of the year is covered by new arrivals.

[6] Average of the one third in the medium profit group.

Source: a) Tanganyika Agricultural Corporation, Reports and Accounts.
b) Cooking, W. P., and R. F. Lord: The Tanganyika Agricultural Corporation's Farming Settlement Scheme; in Tropical Agriculture, Vol. 35, April 1958.

Tobacco in Urambo (Western Region). The consolidation of the groundnut areas in Urambo was first attempted within the framework of private mechanized farms engaged in mixed farming. This proved to be uneconomical. Large farms with the usual crops of the area proved as unsuccessful under private management in Urambo as they did in the case of the state-run farms of Nachingwea. Upon the suggestion of the East African Tobacco Company the cultivation of tobacco was started. A form of settlement came into existence which includes several stages:

Tobacco School. For newcomers there was a tobacco school, which shut down recently. Settlers who had no experience in tobacco cultivation spent their first year here, i.e. they received approx. 2 acres of land and cultivated it under close supervision. Tobacco leaves were delivered to the tobacco shed and dried under supervision.

Small Farms. Those settlers who successfully completed the tobacco school course became tenants of farms of 20—40 acres, on which, at the most, 4 acres could be used for tobacco [1]. Since the closing of the tobacco school in 1962 new settlers are immediately placed on small farms and given especially careful supervision, i.e. 2—3 visits per week. By now a group of experienced tobacco farmers has developed. Thus the tobacco school as the first stage of training is no longer regarded as necessary. New settlers are expected to arrive with a few hundred shillings savings and to build their own house and tobacco sheds.

Medium Farms. Tenants who have successfully operated small farms and have accumulated some savings can progress to larger farms of 150—200 acres, where they are permitted to cultivate tobacco on 15—20 acres.

Large Farms. Most of the land in Urambo is leased to large farms. These holdings comprise 1,200 acres of cleared land, and the right to cultivate tobacco on 60 acres. Until 1962 all of the big tenants were non-Africans. In accordance with the conception of the settlement, two farmers from medium farms were promoted to the tenancy of large farms in 1962. This practice is not to be continued. The new African government does not like the idea of creating "black capitalists" with the help of money provided by the general public or by development aid.

As time passed the number of *large farms* constantly decreased (Table 25). In 1957—58 there were 22 large farms, in 1963 only 11. This is not due to poor profits. Some of the larger tenants left the scheme before independence, taking their savings with them. Potentially interested persons were hesitant to make the necessary investment. In addition, the gradual displacement of large farms by the expansion of smaller ones is in accordance with the original idea.

Medium Farms are operated exclusively by Africans. In 1957—58 there were three, and in 1962 nine of these. The qualification of farmers on medium holdings varies greatly, as is the case in all of the schemes of the TAC. For example, four out of eight farmers on medium-sized holdings achieved success in 1960—61. One of them, who has since progressed to the tenancy of a large farm, sold 90,000 shs worth of tobacco. The other four farmers were not equal to the demands of medium farms, i.e. having to manage numerous wage earners. They returned to the position of a small farmer.

Small Farms are the core of the scheme. Their number increased rapidly. In 1958—59 there were 73, including the tobacco school. By 1962—63 there were already 450, and in 1963—64 750 are expected. The settlement has now gained in popularity — after many years in which it was regarded with caution. Applicants outnumber the available places. In increasing numbers, middle-aged men who enjoy social and professional reputations are applying. The settlement no longer has to depend upon migratory settlers. Applicants become permanent settlers. Since the cleared land suitable for tobacco is already in use, newcomers are willing to take over the hard work of further land-clearance. They have to invest labour and savings in order to construct a tobacco shed. This indicates that Urambo is finally regarded by Africans as a solid proposition.

The explanation is obvious — high cash incomes. Gross income from tobacco fluctuates between 3,000 and 5,000 shs per year. When expenses and the wages for seasonal workers amounting to 1,300 shs are deducted (small farms in Urambo employ some wage labour), a family income of 1,700—3,700 shs remains. In addition, the settlers cover approx. three-fourths of the family's need for foodstuffs by

[1] Only a part of the farmed area is suitable for tobacco. Tobacco should be grown only once in a four-year period. Thus the larger part of the area is kept fallow.

cultivating other crops. The settlers therefore earn 5—10 times the income of the neighbouring peasants without tobacco. Even though the conditions of production are practically equal, incomes fluctuate widely. In 1960—61, 34 out of 128 small farms were not able to achieve a profit (returns from tobacco minus expenses and wages).

Table 25. *Development of Tobacco at Urambo*

Year	Number	Tobacco Acreage per farm	Yield lbs/acre	Price shs/lb	Gross Returns on Tobacco shs/farm	Yield shs/acre
			1. *Large Farms*			
1957—58	22	57	539	3.53	103,000	1,793
1958—59	21	57	654	3.04	113,000	1,984
1959—60	18	61	702	3.16	135,000	2,215
1960—61	11	57	1043	3.11	184,000	3,254
1961—62	14	54	333	3.34	59,000	1,113
1962—63	11	n.a.	n.a.	n.a.	n a.	n.a.
			2. *Medium Farms*			
1957—58	3	13	414	3.19	17,300	1,300
1958—59	6	16	413	2.78	18,200	1,148
1959—60	6	20	529	3.18	33,300	1,682
1960—61	8	22	632	2.94	40,100	1,861
1961—62	7	14	468	3.04	19,300	1,432
1962—63	9	n.a.	n.a.	n.a.	n.a.	n.a.
			3. *Small Farms and School*			
1957—58	—	—	—	—	—	—
1958—59	12 + 61	3.5	397	3.02	4,180	1,195
1959—60	100	4.1	407	3.12	5,190	1,268
1960—61	128 + 99	2.8	625	2.89	5,070	1,812
1961—62	228 + 102	2.2	454	3.10	3,160	1,408
1962—63	450	n.a.	n.a.	n.a.	n.a.	n.a.
1963—64	750[1]	n.a.	n.a.	n.a.	n.a.	n.a.

4. *TAC-Urambo, total*

	Tobacco acreage	Tobacco sales	
1958—59	1633	1,000 lbs	1,000 shs
1959—60	1620	957	2891
1960—61	1430	996	3140
1961—62	1580	1147	3500
		630	2013

[1] Estimate made by scheme's manager.

Source: Tanganyika Agricultural Corporation, Reports and Accounts.

The remaining farms, on the other hand, earned family incomes of 4,000 shs from tobacco alone. The ambitious settler can apparently earn an income which is seldom achieved by skilled labour in towns.

The settlers on small farms till their fields with the hoe. They are responsible for plantig, harvesting, and storage. Instruction is given by settlement officers. Each individual packs and sells his tobacco separately. In this way a direct relationship between the efforts and results of individual farmers is established. It proves to be the settlements' key to success.

The TAC charges a rent of approx. 120 shs per farm. It provides supervision. A settlement officer is available for every 25 settlers. The personnel is made up of African staff members of the TAC who have been with the organization for many years. There are only two Europeans in the scheme, one in the position of a settlement manager and the other as technical adviser. TAC gives a "take-off" loan of about 1,300 shs per farm (mineral fertilizer, insecticides, fungicides, payment for four-months seasonal labour, and transport of firewood for drying the tobacco). TAC also acts as middleman between farmers and monopoly buyers of the East African Tobacco Company. The repayment of loans and interest (5 per cent) is relatively simple, due to the "one channel" marketing system. It is deducted upon payment of tobacco sales.

The fact that there is "close supervision" but no outright compulsion, contributes to the success of the scheme. It almost never happens that a tenant is expelled because he does not farm his holding according to the rules. Indirect pressure is applied. The tenant who neglects his obligations receives no loans. Mineral fertilizers are in this connection a "power key". Without fertilizers tobacco cultivation is not profitable. The fertilizers can be obtained only from the TAC. A special mixture is used. Thus by denying or granting fertilizer loans a benevolent pressure can be exerted. When a tenant is refused a loan the settlement officer informs the "Tenants' Association". Useful in this connection is the tribal tradition, the aspiration to conformity, which appears to be deeply rooted in the African. The tenant who does not follow the rules sets himself aside. "Neighbourly influences" are exerted. The "outsider" either improves or leaves the scheme of his own accord.

The main technical aim is a high yield per acre. Table 25 indicates that the cultivated area per tenant has fallen and yields have risen. The gap between yields per acre on small and on large farms has decreased. The tenants have learned and improved. Experience in Urambo indicates that within the framework of hoe cultivation high yields per acre are much more important than many acres per family. The failure of earlier attempts at motorizing on the one hand, and the success of tobacco cultivation on the other, indicate further that it is not advisable in the early stages of agricultural development to give priority to externally financed motorization. The impetus comes rather from pushing cash crop production of crops bringing high monetary returns per acre. The additional income is by no means completely consumed. Farmers usually save. For the tobacco farmer the objective of saving is the purchase of a tractor. Under these conditions motorization is a result of agricultural development, not the starting point.

With tobacco cultivation the land cleared by the groundnut scheme in Urambo — after unbelievably high losses — is finally finding its feet. The tobacco scheme, however, does not yet pay for itself. Management and supervision necessitate a grant of about 140,000 shs per year. The claim that farmers who earn such above-average incomes as those in Urambo should no longer be subsidized, is justified. An increase in the rents per holding would be the most rational economic way of covering costs. Psychologically this is impracticable. It would endanger the hardly consolidated success. Yet in view of tobacco sales of 2—3 million shs it should be possible to achieve complete cost coverage by levying a turnover tax on tobacco which applies to the scheme only.

The example of Urambo indicates various items which are worthy of imitation. Tobacco is an ideal crop on marginal land. Its production requires little capital investment. The principle of tobacco under "close supervision" can be expanded. Outside of the TAC grounds in Urambo, lots of peasants and land could be associated with the existing nucleus of supervised production. A start has been made at Tumbi, not far from Tabora, where the East African Tobacco Company grants licences for 1—2 acres of tobacco to independent farmers, who have to follow certain rules. It is, however, questionable whether the handing over of the Urambo scheme to an independent co-operative society, planned for 1963—64, is a sensible measure. Management and supervision of such schemes have to come from an external institution.

Ranch and Settlement in Kongwa. Kongwa was originally a farming settlement analogous to that in Nachingwea. The tenants received 10—20 acres of land. The tenant who farmed his land well could receive more. The management of the scheme organized the use of tractors, crop rotation, seeds, fertilizers etc. and deducted costs from crop returns. Farming conditions, however, were marginal for the use of tractors and "close supervision" did not pay. Crop yields (groundnut, maize, castor seed, beans) proved to be low and — more important — unreliable. It was clear after a few years that it would be impossible to offer attractive incomes to tenants and at the same time recover the costs of the scheme.

The successful 90,000 acre ranch of the TAC in Kongwa is in the immediate neighbourhood of the scheme. It has become one of the show pieces of modern agriculture in Tanganyika [1]. It demonstrates that ranching is the most suitable form of land use in Kongwa. The management of the settlement drew the correct conclusions and began gradually to introduce cattle husbandry under "close supervision". There are about 30,000 acres of grassland available.

In the course of time a form of co-operation between the ranch as nucleus and the scheme as a satellite has come into being. The ranch leases to each tenant up to 10 Zebu cows (in calf) at the favourable rate of 30 shs a year in the initial stage, and 40 shs a year in the following years. The calves become the property of the borrower. He can thus build up his own herd of cattle. As soon as his herd reaches 10 heads of cattle (excluding calves) he is obliged to return the cows which he has borrowed. Cattle sales are regulated. The tenants keep at least 5 head and at most 20 head of cattle. Male animals must be sold through the ranch. The female cattle can be used at the discretion of the tenant. The tenant who borrows cows agrees to pay 17 shs per head annually for regular dipping and for amortization of investments in drinking water. The directives of the settlement officers have to be followed. They concern selection of bulls, mating times, castration, grazing rotation, grazing density etc. Settlement officers are helped by the facilities of the ranch. Every service, however, which the ranch provides must be paid for. The annual contract of tenancy is not renewed for those who do not follow the rules. Participants who leave the scheme either on their own, or because they are forced to, must conform to certain rules which ensure that the highly-bred cattle is not lost to the scheme.

The change of emphasis from crop farming with tractors to cattle husbandry is proving advantageous. The number of tenants is increasing rapidly. Fluctuations are small. The Wagogos, a cattle-raising tribe around Kongwa, are especially attracted to the scheme, which makes it possible for them to build up good, privately-owned herds of cattle. In 1956—57 the scheme was started with

[1] The ranch is making profits now, i.e. 10 years after its beginning. No unqualified conclusions should be drawn as to the profitableness of this undertaking. Considerable amounts were invested in cattle, watering places, fencing etc. There are the accumulated costs for personnel in the years in which cattle sales were low; also cleared land was available. When all of these factors are considered, the present profit is probably not sufficient to account for suitable amortization and interest.

300 head of cattle. In 1960—61 there were already 1,500 head (Table 26). If this trend continues cattle husbandry will soon become the major economic activity of the settlement. It is even possible that farming will, in the course of time — when the available tractors are no longer useful — be reduced to a volume which can be accomplished with the use of the hoe.

The settlement does not, of course, achieve the same results as the ranch. There is no fencing: it would be too expensive. Cows, young cattle, and calves are not always separated. There is no night grazing. The cattle is driven daily from the village to the grazing areas. The animals have to cover great distances. The calves do not receive all of the milk. Part of it is used in the household. Yet the most important

Table 26. *Development of Kongwa Settlement*

Year	Number of tenants	Acreage cropped[1]		Cattle		Number of tenants who left scheme at the end of the year
		total	per tenant	TAC[2]	Tenants[3]	
1954—55	19	209	11	n.a.	n.a.	17
1955—56	19	335	18	n.a.	n.a.	8
1956—57	50	1000	20	250	50	9
1957—58	81	1095	13	364	155	13
1958—59	90	1520	16	277	222	12
1959—60	99	1600	16	377	361	16
1960—61	113	960	9	540	1143	5

[1] Excluding fallows.

[2] Cows which the TAC leased to tenants.

[3] Mostly calves and young cattle. The figure for 1960—61 includes 90 cows.

Source: Tanganyika Agricultural Corporation, Reports and Accounts.

problems of cattle husbandry in Tanganyika have been solved: health control, breeding, grazing density and water. There are 90—95 calves to every 100 cows, and 98 per cent of these survive. The use of high quality bulls improves the quality of meat. Sales via the ranch open up better markets. The tenants receive 40 cts/lb liveweight for four-year-old steers, compared to prices of 20—30 cts in other areas.

Building up capital in the form of cattle facilitates "close supervision". As cattle must be sold via the ranch, and as such sales can be more easily controlled than in the case of groundnuts, the management of the settlement disposes of an effective instrument for loan repayments. The debts incurred in the course of the year are settled at the time the sales are made.

Matengoro, an Attempt to repeat the Kongwa Pattern. The Kongwa cattle settlement operates under favourable conditions. Cleared land is available, both for crop farming and grazing. Roads, installations and buildings had already been constructed by the groundnut scheme. The TAC made additional investments. Crop farming carried the scheme over the early stage of building up the herds. Such considerations aside, the essential features of the Kongwa Settlement may serve as an example for other schemes.

Cattle husbandry under "close supervision" requires relatively little capital. The most important factors of production — cattle and grazing — are available. There are additional costs for the construction of watering and dipping facilities. In many places such facilities have already been constructed, but are not used. The decisive element of the Kongwa model — a better combination of the available factors of production under close supervision — is relatively inexpensive, especially in cases where nucleus ranches exist.

An attempt to reduce the Kongwa model to terms which would allow it to be copied in Tanganyika was undertaken in 1962 in Matengoro, a few miles away from Kongwa. Local Authorities offered 20,000 acres of bush and grass. Under optimum conditions the area can support approx. 3,000 head of cattle, including 800 cows. The TAC places tenants on the land. They are to bring at least five head of cattle with them and are not permitted to keep more than twenty head. A Settlement Officer supervises health, breeding, grazing density and rotation. His directives have to be followed. All expenses, plus 24 shs general administrative fee per tenant, must be paid. If it proves possible to persuade the tenants to bring their own cattle, the expensive building up of cattle herds, which was effected in Kongwa by the subsidized leasing of cattle, can be achieved in a much shorter period.

Mechanized farming is to play no part in the scheme, cattle will be the "cash crop". The participants are to raise crops for home-consumption under the usual system of shifting cultivation. There are some advantages to this. The scheme includes extensive areas of flat grassland (mbuga), which provide abundant fodder in the dry season. The hills on which the cattle should graze in the rainy season are overgrown with bush. Shifting cultivation would serve as a method of gradual bush-clearance. The period of crop farming would not be followed by the regeneration of bush, as is usually the case, but by the transition to grassland, facilitated by heavy grazing, clearing of bush, and planned burning every four years.

The settlement at Matengoro is relatively inexpensive. The initial investment for 1) planning, 2) demarcation of the borders, 3) construction of feeder roads, 4) water supply, 5) dipping and 6) housing for the settlement officer, amounts to approx. £ 3,600 [1]. The current costs of supervision are £660 annually. If the original investment is amortized in 10 years the annual costs will amount to £ 1,000, i.e. 20,000 shs. On the other hand, a fully stocked scheme which involves 800 cows, can make annual sales of 700 head of cattle, each weighing an average of 700 lbs, i.e. at the price of 35 cts per lb liveweight, sales valued at 171,500 shs. So the input-output relationship is likely to be very favourable.

Whether the system of cattle husbandry under "close supervision" can be realized remains to be seen. Some difficulties are already apparent. The tenants are not always willing to bring their own cattle with them. They hope that cattle will be leased from the ranch and that they will be able to leave their own cattle outside the scheme, in the care of relatives. Tenants request that their arable land be tilled by tractors. Both of these measures would make the project unduly expensive. It has been demonstrated often enough that under these conditions the use of tractors is not profitable. The tenants' failure to bring their cattle with them means that herds are built up slowly and that, therefore, high costs accumulate for personnel, amortization and interest. Problematic is the fact that the area is still legally owned by the Local Authority. The TAC is only acting as managing agent. On the basis of previous experiences in Tanganyika it is doubtful, if not improbable, that the TAC will receive sufficient support from the Local Authorities if difficulties arise due to implementation of the rules or to the expulsion of unco operative tenants. In Kenya, almost all grazing schemes which took in more than 2.5 million acres of land have failed because of such difficulties.

[1] Several hundred emergency workers were employed in Matengoro for bush clearance. They were fed from US-supplied maize. The costs amount to several thousand pounds. Little was accomplished. This outlay is not economically justified.

In spite of these considerations the TAC, by the example of Kongwa and Matengoro, has provided an outline for the reorganization of cattle husbandry in Tanganyika which is psychologically acceptable in many areas and which can be profitable for the Treasury.

In areas of low rainfall cattle husbandry can in this way become a "cash crop", while hoe cultivation can continue to provide subsistence. Individual use of grazing areas is not feasible. A division among numerous owners would give rise to insurmountable problems of demarcation (fencing and hedges) and water supplies. Cattle must migrate according to the season. An economic ranch must dispose of some thousand acres. Thus the only alternative lies between large ranches and communal use. In view of the absence of professional knowledge, and the state of mind of the herdsmen, independent cattle co-operatives are not a realistic proposition. There is thus no other alternative than supervised ranching as demonstrated by the TAC.

b) Plantation Crops under "Close Supervision"

Another principle of British agricultural development policy in Tanganyika was — beside the approaches to land settlement already mentioned — to introduce peasants to the production of crops grown up to then on plantations only, while using the plantations as development nuclei. Of course, plantations have frequently functioned as development centres for neighbouring peasants. The plantation economy has contributed to the spread of skills, new seeds and better cattle. In the late fifties development went further. Peasants were to participate in producing plantation cash crops. This was made possible by placing their production under "close supervision" and by processing their production in plantationowned factories.

α) Peasant Tea

Examples in Kenya and Uganda show that peasants can cultivate tea and achieve economic success with it. In Kenya there are about 6,000 acres of peasant tea. The farmers pluck the leaves at fixed times and transport them to collection points. Inspectors examine the crop and organize transport to the factory. The advantage of peasant tea is that significant capital formation in the form of tea planting can be achieved with very little outlay except family labour. The establishment of one acre of tea costs about £ 300, of which not more than £ 40—60 are expenses. The rest corresponds to costs which can be carried by family labour. Peasant tea was therefore, introduced in the Usambara Mts. and in the Southern Highlands. Private plantations agreed to buy leaves of good quality from the farmers. The costs of processing are deducted from the price.

The procedure is as follows: Extension officers inform the peasants in suitable areas of the possibility of earning additional income through tea. Each field of each farmer interested is examined and tested for suitability. Participants may be granted a loan of about £ 13. This is sufficient to buy stumps for 1/4 acre of tea. Planting is done under close supervision. In some cases assistant field officers and instructors mark off each terrace and position for each plant. If the work is done correctly the participants receive a licence to cultivate 1/4 acre of tea only. If this is successfully cultivated they are allowed to plant up to 1 acre. Negligent farmers lose their licenses. The repayment of loans is controlled. The leaves can be sold only to the factory. Debts are deducted from the price paid.

The income possibilities from tea are remarkable. The first full crop can be expected after four years. Then a yield of 3,000 lbs of green leaves per acre with a price of 40 cts/lb brings a gross return per acre of 1,200 shs. Peasants are interested. The Usambara tea scheme started in 1961. Within 18 months 75 acres were planted. In Lupembe, Southern Highlands, 150 acres have been planted in two years. In Tukuyu it can be observed that the same peasants who neglect their coffee carefully husband their tea-plots, although there is no harvest before 3 or 4 years [1].

The tea schemes raise difficult questions of organization and supervision. Experience shows that — as time passes — the quality of work tends to deteriorate. Initial standards are seldom kept. Success or failure depends on yields per acre. Peasant tea in India and Ceylon yields per acre only about 1/4 of what is harvested on the estates. Yields in Kenya are good [2]. Cultivation is, however, closely supervised. Warnings are given to those who ignore the rules of good husbandry. Repeated offences lead to the suspension of the licence or to legal action. In Tanganyika supervision is not as strict. Furthermore, the high cost of personnel must be considered. The 75 acres in the Usambara Mts. mentioned above are the result of two year's work by one officer, 4 assistant field officers and additional instructors. Yet Bridger, who considers all these factors, assumes that total input, including cost of personnel, subsidized loans and subsidized transport of leaves, stands in a favourable relationship to the additional tax revenues which can be expected.

According to Bridger the economic balance for 300 acres of tea on a small farm can be calculated as follows [3]:

Input:	Management and supervision, recurrent costs plus costs of the first 6 years, distributed over the next 20 years, annually	43,100 shs
	Lost loans for stumps and mineral fertilizer	20,100 shs
	Annual sum	63,200 shs

[1] Targets for 1968 run as follows: Usambara — 848 acres, Tukuyu — 720 acres, Lupembe — 682 acres and Bukoba — 1,055 acres.

[2] The same is true of Uganda.

[3] Bridger, G. A.: Peasant Tea Production in Tanganyika. ECA/FAO, April 1961, Mimeo.

Output:	1. Additional farm income, expenses deducted		
	(1,140 shs/acre)	342,000 shs	
	tax revenues 12 per cent thereof		41,040 shs
	2. Additional income of the processors . . .	90,000 shs	
	tax revenues 20 per cent thereof		18,000 shs
	3. "Primary income" total	432,000 shs	
	4. "Secondary income" i.e. 1.5. times (3) . .	648,000 shs	
	tax revenues at 17 per cent		110,160 shs
	Total additional income	1080,000 shs	
	Total additional taxes		169,200 shs

If the above calculation is approx. correct, then peasant tea is highly profitable for the Treasury. An annual input of 63,200 shs for 300 acres of tea bring in effect additional tax revenues of 169,200 shs.

β) African Sisal

Efforts to obtain peasant participation in plantation crops are not restricted to tea. TAC settlements in Kilombero valley (Kiberege strip) grow sugar-cane. The production of cocoa and rubber is scheduled. As to sisal, however, no actual beginnings have been made. There is no peasant sisal under "close supervision". The question how to get more peasant sisal is under discussion. The approach can be considered typical of economic policy shortly before and after independence. The following considerations apply:

When the most important industry of a country — in Tanganyika the production of sisal — is exclusively in the hands of foreigners, i.e. Europeans and Asians, who operate estates and employ a large number of wageworkers, it constitutes a dangerous situation, at least in the present social and political situation. Arguments are bound to arise: foreigners try to keep a monopoly of the most important cash crop, profits are transferred, economic power remains in the hands of plantation owners whose loyalties lie outside Tanganyika, etc. This, in turn, is likely to create economic instability. There will be less investment in sisal. This will bring stagnation to an industry which is essential to the nation and to its economy. There are, in addition, other considerations:

- Wages of sisal workers have almost doubled in 1½ years. In 1960 they amounted to 46 shs per month plus free housing and food. In 1962 they reached 126 shs without food. Higher wages can induce "marginal" producers on soil of inferior quality, in dry areas, or those with little capital equipment etc. to stop production. This tendency is presently being held back by high sisal prices. It would at once become apparent, should prices fall to the level of 1957—58. Peasant sisal is not touched by higher wages. In this connection, the long range prospect opens up of a division of labour between peasants who produce sisal leaves and the processing factories.
- Also, in some of the best developed areas there is no room for additional estate sisal. Areas which have been excluded from use by Africans are already planted.

On the other hand, there are extensive neighbouring areas suitable for sisal which are available only to the peasants. It will be necessary to develop a form of co-operation which is attractive to peasants, so that the sisal factories can be supplied with leaves from these areas.

African sisal may be cultivated using the following methods:

— State-owned *estates,* run by an organization like the TAC. Economic results will probably be better if a private firm is used as "managing agent". The private firm thus administers public — that is African — capital. It can be asked to train and employ African staff. As to finance, the establishment of joint enterprises with state and private capital is feasible.
— *Settlements under "close supervision".* Participants will have to do the work in a given area and they will be paid according to the return. The position of the participants might be similar to those of share croppers elsewhere.
— Cultivation of *peasant sisal around existing plantations.* The leaf is sold to processing factories. Sisal may be planted by individuals or communally.
— *Hedge-sisal,* i.e. peasants plant hedges on their land. They cut the leaf, decorticate it and sell the fibre.

The establishment of estates is undoubtedly technically and economically feasible and probably profitable at existing prices. The initial investment, however, is extremely high. Approx. £ 250,000 is required for one sisal estate. The government is hardly in a position to finance many of them. Leaving investment exclusively to private capital does not solve the problem of Africanization.

Sisal settlements under "close supervision" create a number of problems to which one cannot find a reasonable solution. There is not the opportunity — as with tea — that most of the initial investments will be made by family labour. Sisal grows in a cycle which extends over 8—12 years. Huge amounts of fresh leaves have to be transported. Consequently, large fields must be established and cut in rotation. Participation of settlers would require everybody to farm some sisal in all age groups, in order to provide an annual income. This, in turn, would lead to transport problems which can hardly be solved economically. There is only one aspect worth trying: Combining plantation sisal with supervised workers' cattle. Sisal workers might be allowed to graze their cows between growing sisal. These cows could be put — as in the Kongwa settlement — under the supervision of the scheme's management. Whether this would work, however, remains open to some doubt. All in all, plantation sisal is not a crop for settlements under "close supervision".

Peasant sisal around existing plantations has a somewhat better chance. Border plantings are an interesting proposition for some commercially minded medium farmers, who are able to handle tractors, and who might develop into satellites of an estate. Thus border planting is the answer in some specific cases. It is not a general answer to the problem. There are cases of idle capacity in leaf-processing factories not too distant from peasant land. In 1962, 21 cases of border planting were registered in Tanga Region. They cover up to 20 acres of sisal each. Work was done communally, initially with much enthusiasm. As time passes, however, weeding is neglected. Interest slowly dwindles. The reason is clear: sisal returns per acre are low compared to other cash crops such as tea or sugar cane. Plantations pay 15 shs cbm/leaf. A yield of 500 cbm/leaf/ha within the 10 years life cycle of a sisal plant, corresponds therefore to an annual income of 750 shs/ha or 300 shs/acre[1]. Thus the gross return on sisal is hardly higher than on a good crop of maize or manioc. Plantation sisal is not an interesting crop for a man who works with the hoe.

Therefore the most promising approach to the Africanization of sisal is the *improvement of hedge sisal.* Hedge sisal is already an important crop in Sukumaland. Production fluctuates widely. Some twenty brushing factories processed 200 tons of fibre in 1957 and 15,000 tons in 1962. The quality is poor; the hedges are neglected; the leaves are wrongly cut. All in all, harvested hedges yield hardly half of their potential. In addition, there are innumerable uncut hedges. It is obvious that improvement in hedge husbandry, exploitation of unused hedges and the planting of new ones could bring about rapid and cheap increases in hedge sisal production. In a very short time a volume of 45,000 tons and more could be reached. Hedge sisal could become a major industry in Tanganyika, as it is in Brazil. The difficulty is decortication. Hand decortication takes too much time and produces poor qualities. The key to proper peasant sisal, therefore, lies in the development of small, power driven decorticators, machines that could be transported to the hedges. Decortication could be done by contractors on the spot. Whether it will be possible to obtain acceptable fibres without the intensive washing process, which up to now ties sisal production to large estates with dams and factories, remains to be seen. If present efforts in this direction succeed, peasant sisal will gain tremendously in its competitiveness with plantation sisal.

c) Irrigation and "Close Supervision"

The British administration in Tanganyika was mainly interested in agricultural extension and settlements on non-irrigated land. Shortly before independence additional interest was attached to peasant production of plantation crops. There was hardly any noteworthy interest in irrigation. Substantial sums were invested in providing drinking water for men and

[1] The sisal industry measures in hectares.

animals. This led to the construction of several small dams, with some minor irrigation farming attached to them. Not a single larger irrigation scheme came into existence under British rule, except the trial project at Mbarali in the Southern Highlands and some private undertakings in connection with sugar cane production.

The following explanations are put forward:

— An unfavourable input — output relation can be expected. The projects in question are for the most part in areas which are thinly populated. In addition to the actual costs of the project, high costs for infrastructure, flood control, population resettlement etc. are to be expected. On the other hand, prices are low. There is no internal market of significant buying-power. Irrigation schemes will have to produce export crops. High transport costs to ports must be taken into account.
— One major irrigation scheme alone does not suffice. The economic usefulness of irrigation in the Rufiji Basin depends upon the construction of a system of dams, irrigation facilities and measures for flood control. This again exceeds the country's capacity. It raises questions of the availability of capital and labour, and of the utilization of power and production, which greatly exceed Tanganyika's economic capacity.
— The decisive argument is that a better utilization of available funds is seen in the promotion of numerous small agricultural projects.

The work on larger irrigation schemes did not pass the preparatory stages. Plans and studies were made for the use of the Pangani (Tanga Region) and the Ruvu (Eastern Region). Well known is a study on the possibilities of irrigation within the Rufiji Basin which was financed from British aid and carried out by the FAO (see appendix B). So far, the experience gained from the Mbarali trial scheme has contributed to fortify British prejudice against irrigation in Tanganyika.

Work at Mbarali began in 1958. For 1960 a cropped area of 2,250 acres was scheduled, and the actual acreage achieved up to 1963 amounted to 500 acres; 250 acres are distributed among the settlers at 5 acres each. The rest is farmed by the scheme management. The scheme is more costly than was expected. Gross investment is likely to exceed 2,000 shs/acre. There is no high-yielding cash crop. Cotton, which could justify high investments, cannot be grown. Mbarali is situated in a belt which is intended to protect the production of cotton in other regions from insects coming from Mozambique. The crops grown (maize, groundnuts, and beans) hardly return more than 400 shs/acre. Thus gross returns remain around 20 per cent of the gross investments, while 33.3 per cent is the key figure for profitable irrigations.

The gross returns per settlement will reach 2,000 shs in good years. From this, 800 shs have to be paid for water and amortization. An additional 750 shs are charged for tractor ploughing and other services. What remains is definitely not enough to attract ambitious settlers. There is no production under "close supervision". Decisions as to the use of water, choice of crops etc. are made by farmers who have no irrigation experience whatsoever. Thus Mbarali as it was in 1963 is an example of how not to run irrigation schemes. The main reason is lack of initial planning. Good water and good land — both are available at Mbarali — are not sufficient criteria for investment in irrigation.

Mbarali, as well as experience elsewhere in Tanganyika, indicates that the problems of irrigation are not primarily of a technical nature. The economic and institutional aspects are much more involved. Successful irrigation requires land consolidation, planning of land use and water distribution. High yielding cash crops are required. The burden of amortization can only be carried if all the rules of modern farming and careful husbandry are applied. The principles of amortization and interest are not usually understood by the farmers. Administrations are unable to make appropriate payments. Most farmers believe that water, a free gift of nature, should be provided free of charge. An example:

In 1955 a project to settle 66 tenants on holdings of 7 acres was started near a heavily populated area. At least 100,000 shs were invested in irrigation facilities. The plan was that the tenants should make an agreement with the owners, i.e. in this case with the Local Authorities, to develop the holdings and to farm them properly. Annual rents were put at 250 shs per tenant. Allocation of water and the maintenance of the canals was to be organized by a committee of tenants. The plan failed. Some of the holdings remained idle. Canals were not maintained. Water was lost. The system of canals and sluices proved inappropriate. No one felt responsible. The Local Authorities had other things to do. The local Agricultural Officer did not have sufficient powers. The committee for the organization of water supplies was never effective. The tenants did not live on their holdings. They mistrusted the status of a tenant. Rents were not paid. In short, the experiment is, after 6 years, a picture of complete disorganization.

The experience here and in other areas shows that investment for irrigation is rational only if:

- Thorough preparatory work is carried out, including levelling of the ground and laying out of each field. The tenant should live on his holding. If he does not have any starting capital, loans must be provided.
- Continuous controls ensure that the water is properly used and that technical difficulties are ironed out in the early stages.
- Farmers are willing to pay amortization and interest. A rate of 25—33.3 per cent of the gross returns may be considered adequate. These payments can probably be achieved only if sales are made obligatory via a sellers' co-operative and if these can be controlled. The debts can then be collected from crop returns.
- Crop rotation, manuring, use of insecticides, husbandry etc. are closely supervised.
- A responsible settlement organization, in the case of Tanganyika the TAC, takes over the management of the scheme. The settlement organization must have enough power to be able to expel those tenants who are not willing to co-operate.

If the farmers are not willing to fulfil these unavoidable conditions for successful irrigation, settlement should be postponed. It is undoubtedly better to resort to large scale farming than to establish settlers who consume funds made available by the general public or by development aid.

The British administration in the late fifties was not able or willing to enforce the above-mentioned conditions for successful irrigation. The administration was not even able to raise the running costs of some smaller irrigation schemes. In such a situation, investment in irrigation amounts to subsidizing those who are already in a preferential position, thanks to the fact that irrigation land has been allotted to them. It is understandable that a British administration in a UN territory shied away from solving tricky irrigation questions, particularly since it was assumed that the extension approach is much more economic.

Because it seemed difficult and unwise to implement "planned and supervised" irrigation, efforts in the sphere of water use remained restricted to the improvement of existing facilities. Scattered throughout Tanganyika are some 10,000 acres of peasant irrigation. The furrows of Mt. Kilimanjaro and the Usambara Mts. which carry water over many miles are well known. Agricultural administration tried to foster local interest in the works. Technical advice, cement and grants were occasionally provided.

d) Summary of Production under "Close Supervision"

The British administration did not have enough time, or started too late. There are no established patterns yet for successful production under close supervision. Efforts in irrigation led to nothing but failures. As for tobacco, cattle and sugar cane, the schemes are still too expensive to be profitable in Treasury terms. The approach, however, is promising and the principle is right. It is as follows: an increase in cash crop production by new combinations of the existing resources of labour, man, and capital. Close supervision involves the application of technical knowledge, which the farmers cannot do by themselves. The power of the state as landlord provides the necessary pressure in order to induce participants to follow rules. Actually all schemes require some additional capital investment. It plays, however, a secondary role (see tea, cattle, tobacco). Basically, production under close supervision is a way to increase production through the introduction of innovations rather than through capital investment.

It must be admitted, however, that production under close supervision is technically not as efficient as estate production. On the other hand it lies far above ordinary peasant levels. Of special advantage is the fact that farmers obtain the chance to effect considerable capital formation through land clearance, building up of high quality cattle herds, establishment of tree crops and investment of savings to productive ends. Production under close supervision is probably a cheaper and more rapid way towards capital formation in agriculture than the establishment of large state farms.

All in all, basic considerations and current experiences with production under close supervision are promising, if — and this must be strongly emphasized — the following conditions are taken into account:

— Production under close supervision is not worthwhile under marginal conditions. Costs are high. Trained and talented personnel, i.e. expensive personnel, is necessary. Economic success depends upon concentrating on profitable, large schemes with high-value cash crops. The schemes have to be simple. Monoculture, or restricting supervision to one crop, is advisable. As a rule the crop should be one which produces high gross yields per acre (tobacco, cotton, sugar cane, tea). There are exceptions. Ranching probably also belongs to those economic activities in which supervision would pay.

— The necessity of restricting close supervision to some profitable schemes leads to a two-way approach in production development: Settlements on land of low fertility do not justify close supervision. Such land should not be made the target for ambitious schemes. The traditional, unregulated occupation of land by the farmers themselves is the most suitable economic measure. On the other hand, it may be worthwhile thoroughly to plan settlements in areas of high potential and to place them under close supervision. Small irrigations are not suitable places for the "leap" into modern farming under close supervision. They can gradually be improved. Counselling, technical aid and grants are appropriate here. Larger scale irrigations, however, require close supervision.

— As far as settlement schemes are concerned, BRIDGER's statements are applicable[1]:
"It is a widespread failure to construct settlements in thinly populated areas and to allow the population of these areas to participate in the settlement. The people in thinly populated areas already earn a moderate income; the desire to work harder in order to earn a higher income — which is for many of questionable use — is usually not present. Discipline is rejected and the farmer returns at the first opportunity to his own farm. Settlement projects should therefore be made available to the inhabitants of thickly populated areas. Attempts to construct settlements for people of thinly populated areas fail, or prove to be much too expensive."

— Thorough preparation is necessary. Schemes which are begun with too little capital and incomplete equipment accumulate high costs for personnel before additional revenues can be expected.

— The responsibility for schemes under close supervision must be carried by an external institution, for example the TAC. It may work with local committees. The competence of a local institution, for example the Local Authorities or local production co-operatives, gives rise to almost insurmountable human and political complications. Scheme managers must be able to operate as independently as those of large private enterprises. They must have enough power to be able to expel those who do not follow rules. Hence they need political backing.

[1] BRIDGER, G. A.: ECA/FAO Joint Commission for Africa.

— The participants in a scheme under close supervision should be placed in a position between "incentive" and "pressure". Encouragement alone is of as little use as are regulations by themselves. The most important incentive is a high cash income. It does not suffice to reckon with family incomes of 1,000 shs or 1,500 shs. The success of the scheme depends largely on getting above-overage farmers. Their income — before entering the scheme — is already much higher than the average. It will probably be wise to provide for family incomes of at least 2,000 shs and 3,000 shs. Other incentives are: the chance to build up wealth through cattle and tree crops, a ladder of economic and social progress for successful farmers, the organization of social services such as drinking water, community houses, schools, dispensaries, etc. The construction of houses is generally not necessary. The attempt to encourage by making "gifts", i.e. by the free allocation of seeds, tools, tractor services, etc., is a danger to the economic receptivity of the farmers. It stifles the initiative for self-help.

— Furthermore, it seems wise to plan schemes in such a way that the participants have scope to employ seasonal wage labour. It can be observed everywhere in Tanganyika where peasant farms are advancing, that seasonal wage labour is employed by the more active and ambitious farmers. Their talent for commercial production is thus making use of people who are unable to "employ themselves" and who cannot find employment in the towns and on the estates. Devising a scheme in such a way that participants must employ others means a broader application of the scarce factor of production: technical know-how. Technical knowledge thus spreads from scheme officers to participants and from participants to employed wage labour.

— In addition to "encouragement", management should have an instrument of enforcement at its disposal, with which it can put "pressure" on hesitant or lazy farmers and those unwilling to make their payments. Such "instruments" of pressure might be: contracts of tenure, licences which can be revoked, or obligatory sale of production through a co-operative or the management. This ensures that outstanding debts are collected. Loans are necessary, not only in order to equip the tenant adequately, but also because they create a situation of dependence. Since in this case farmers are dependent on a "benevolent landlord", such dependence is not harmful. The opposite is true. Dependence leads to a situation where the scheme management can exert pressure in order to enforce good husbandry which, in turn, is of benefit to the dependent person.

— In order to be successful, production under close supervision should be affiliated to a nucleus estate or to a processing industry. If privately-owned large farms are not available, or cannot take on this responsibility, then state-owned farms must be established. Their service should be given to the settlers at full cost.

— Arrangements between the settlers on the one hand, and the estate or the factory on the other, are often plagued by difficulties. Disagreements about quality, prices and deadlines for services are to be expected. Political considerations may also play a part, especially where private estates (tea and sisal) are concerned. In such cases there is probably no other alternative than to provide for governmental agencies which can arbitrate in such disputes.

In its final stages the British administration was not able to enforce these conditions for successful production under close supervision. Its major concern was the preservation of peaceful conditions among the peasants. Yet the experiments made have opened up new avenues of approach. The British administration has assembled capital in the form of experience which the new administration can utilize. With the TAC it has passed on an instrument well adapted for further nucleus estates and settlements.

3. Market Co-operatives[1]

Another essential feature of British development policies in the late fifties — supplementing agricultural extension and production under close supervision — was the support given to African initiative in establishing sellers' co-operatives. Already in the twenties Chagga coffee farmers organized the Kilimanjaro Native Planters' Association. This organization was the predecessor of the KNCU, which was registered as a co-operative in 1932 (see Table 27). Other co-operatives also began in the 1930's. However, up to 1950, growth and success were largely limited to the KNCU. Membership, acreage and sales have consistently increased. Meanwhile the business volume of the KNCU includes one-third of Tanganyika's coffee, valued at more than £ 2 million.

Rapid progress was achieved in the decade 1950—1960. The growing commercialization of peasant farming, the increasing social and political consciousness of country people and the systematic promotion and supervision of the "Departments of Co-operative Development" were the main factors which led to success. In 1949 there were 79 co-operatives, in 1960 there were 691, and in 1963 more than 1,000. Of greatest significance are the large coffee co-operatives in Moshi (KNCU, TACTA) and in Bukoba (BNCU) and the cotton co-operatives in Sukumaland, which are incorporated in the "Victoria Federation of Co-operative Unions" (VFCU). These four groups handle an estimated 80—90 per cent of the total co-operative sales. In 1963 the products sold were valued at a £ 14 million.

---

[1] This chapter about co-operative development has been kept short. Much has been written about co-operatives in Tanganyika. See bibliography.

Practically all peasant production of coffee, cotton and pyrethrum is handled by co-operatives. Co-operatives account for about $^{1}/_{3}$ of the total agricultural exports. Co-operatives also invest in the processing of coffee and cotton (2 plants for the processing of coffee, and 6 ginneries). All of the co-operatives in Tanganyika are incorporated in a Co-operative Union. In 1962 a Co-operative Bank was opened.

Table 27. *Growth of the Kilimanjaro Native Co-operative Union Ltd.*[1]

Period (June – July)	Number of coffee growers in thousands[2]	Acreage[3] in thousands	Green Coffee in tons	Crop value	
				in 1,000 £[4]	£/ton
1923/24—27/28	5,5	2,2	98	n.a.	n.a.
1928/29—32/33	10,7	4,5	600	n.a.	n.a.
1933/34—37/38	21,3	10,1	1073	40	40
1938/39—42/43	26,7	15,8	2199	96	44
1943/44—47/48	29,8	17,9	2539	282	109
1948/49—52/53	33,0	23,5	3730	925	253
1953/54—57/58	39,0	29,2	5672	2637	460
1958/59—60/61	44,8	33,7	7119	2164	306

[1] Known as the Kilimanjaro Native Planters' Association until 1932.

[2] Because of obligatory sales these figures are identical with the number of suppliers of the KNCU.

[3] Estimated figures.

[4] F.O.R. Moshi.

Source: Information from the KNCU, July 1, 1961.

Tables 28 and 29 contain expenses incurred by the coffee co-operatives, the cotton co-operatives and the "Lint & Seed Marketing Board". Primary societies buy the products from the producer. The unions, in one case the KNCU and in the other the VFCU, coordinate the work of primary societies and pass the products on for processing. In both tables the high amounts which the unions pay to the Local Authorities should be noted. The Chagga Council receives 3.1 per cent of the coffee returns, the comparable administrative organs in Sukumaland receive 3.6 per cent from the sale of cotton. These sales taxes are used principally to finance communal projects such as streets, schools, hospitals, drinking water etc.

In a number of cases the British administration was quite successful in keeping a balance between aid and supervision on the one hand, and creating an atmosphere favourable to self-help on the other. Co-operatives were not established by administrative directives, but local initiative was helped[1]. This help in the early stages was indispensable. There was no trained personnel. Book-keeping is a particular problem. The Department

[1] This is mainly true for the late fifties. In earlier years British administration took a rather sceptical attitude.

of Co-operative Development provided recently organized co-operatives with trained staff. The government gave decisive support by making the sale of coffee, cotton, pyrethrum to co-operatives obligatory. The co-operatives are usually not equal to the competition of Indian traders. Obliga-

Table 28. *Expenses of the Kilimanjaro Native Co-operative Union* in shs per bag parchment coffee

Average price per cwt[1]			232.49 shs
Expenses			
1. Twine, labels and brushes	0.36		
2. Transport to the curing works	1.34		
3. Handling, curing and hand picking	5.85		
4. Sisal and jute bags for transport	2.41		
5. Warehousing and sampling charges		9.96	
	0.14		
6. Liquoring	0.06		
7. Auction expenses	0.10		
8. Brokerage	0.59		
		0.89	
9. Interest and bank charges	1.46		
10. Insurance	0.25		
11. Supervision fees	0.08		
12. Maintenance of scales	0.13		
13. Printing and stationary	0.06		
14. Sundries	0.34		
15. Losses	1.13		
16. Tanganyika Independence Celebr.	0.07		
		3.52	
17. KNCU, Commission at 2½ per cent	5.81		
18. KNCU, levy	3.36		
19. Chagga Council Native Treas. Levy	7.41		
20. Societies' levies (average)	8.88		
21. Societies' excess and ind. surplus	1.39		
		26.85	
Total expenses			41.22
Average price to growers per 112 lbs.		191.27	
Average price to growers per 1 lb.		1.71	

[1] There is a loss of 18.86 per cent in weight of the delivered product. For this reason 232.49 shs per cwt of parchment coffee corresponds to the price of 287.00 shs of processed coffee at the auction.

Source: KNCU Circular No. 577 (May 21, 1962).

tory sales are necessary if they are to succeed. In the short run, traders may give better service to the producer than co-operatives. Traders have no monopoly. Competition among them is keen. They have long-established experience in commerce. However, the co-operatives are an indigenous

African institution. Their members learn through failures, set-backs and difficulties, and gain commercial knowledge and experience in dealing with economic power. In the long run, this advantage is certainly more important than any possible losses due to lower efficiency. Obligatory sales may be regarded as an unavoidable protective measure for the emerging co-operative movement.

Table 29. *Expenses of Cotton Marketing*
Lint & Seed Marketing Board, 1960/61

Price for cotton (lint) and cotton seed computed per lb. of seed cotton	75	cts.
Expenses		
1. Primary Society (buying, packing, seed distribution)	5	cts.
2. VFCU (Transport to ginneries)	1	,,
3. Co-operative Union	1	,,
4. Native Authority Levy (streets, schools, hospitals)	2	,,
5. Transport costs	2	,,
6. Ginning expenses	10	,,
7. Handling and sales of cotton seed	2	,,
8. Seed distribution[1]	2	,,
9. Research, cotton development	1.5	,,
10. Warehousing	0.5	,,
11. Lint & Seed Marketing Board (Administration, sales, transport, insecticides for seeds, inspection of ginneries)	0.5	,,
Total expenses	27.5	cts.
There remain for the grower per lb.	47.5	,,
The grower is paid per lb.	55.0	,,
Subsidy per lb.	7.5	,,

[1] The cotton growers receive free seeds.

Source: Lint & Seed Marketing Board. Report for the Year ending June 1961, p. 20.

Co-operative development was helped by marketing boards. This is especially true in the case of cotton. Co-operatives are responsible for only the first phase of the marketing process, the purchases from the producers and the transport to the ginneries. From there on the Lint & Seed Marketing Board is responsible. It takes over the difficult task of export sales. The Lint & Seed Marketing Board also has continuous insight into the work of the co-operatives and can help when necessary. Coffee producers who are not members of the KNCU are incorporated in the Tanganyika Co-operative Trading Agency (TACTA), and thus make use of a private firm as managing agent. The KNCU owns a coffee processing factory together with the organization of coffee-estates — the Tanganyika Coffee Growers' Association. In this respect an exemplary and mutually advantageous co-operation between co-operatives and private firms and between Africans, Europeans and Indians has been established.

Co-operatives are the only indigenous economic institution of the rural population of Tanganyika. Because of obligatory sales they have considerable power. They can exert significant influence on production, not only by their price policy, but also by actively promoting the introduction of innovations. They can play an important role by supplying small loans and collecting repayments, thanks to their monopoly position. These opportunities are little used. Considerable sums which the co-operatives took in, have been used principally for conspicuous buildings. This display of the newly-won status certainly has a psychological value. Members, however, would be better served if production and productivity were promoted, for exemple by giving grants to agricultural extension work. The co-operatives sometimes operate nurseries, give loans and sell implements. They finance the application of insecticides which is organized by the local Agricultural Officer. Such measures, however, are due more to government insistence than to the initiative of the co-operatives themselves. Altogether, the co-operatives are not very active in this respect. They are apparently hesitant to complicate their work by varying their services. They are reluctant to take on tasks which are connected with unpopular measures, for example, the collection of loans. They prefer to have satisfied farmers. This attitude is understandable in view of the necessity to consolidate the current successes. Yet the co-operatives represent an organizational set-up which will in all probability have to deal with such questions in the future.

4. Summarizing Agricultural Development Efforts in the last Decade of British Administration

Tanganyika's economy would look different today if as much had been done in the decades between 1920 and 1950 as in the last ten years prior to independence. The years between 1950 and 1960 are characterized by remarkable progress, particularly in the realm of peasant production. Within 10 years peasant production became the main industry of the country, in terms of total output as well as of market and exports.

This is certainly not due only to British efforts. It indicates a rapid social change among the peasants. Without the services provided by the British, however, existing inclinations to produce for the market would hardly have led to the present volume. Government efforts, and particularly channelling these efforts into a reasonable three-way approach:

— improvement of existing peasant farms through the persistent persuasion of agricultural extension workers,
— transformation of farming in special schemes under close supervision,
— support for local initiative in co-operative marketing

has undoubtedly contributed to the decade of success.

There was no lack of difficulties and failures. Most of the government's agricultural personnel spend most of the time on general administration and — as is the case everywhere — on lots of red tape. Colonial administrations run into serious problems. College-trained agriculturists are expensive, particularly in tropical countries. Extension work is hampered because the farms are small and distances great. Officers have to go on leave every 2$^1/_2$ years. Work is interrupted. On their return the officers are transferred to other locations. There is little continuity in supervising subordinate personnel. Changing officers frequently means a change in work and this rightly so, because abilities differ. 2$^1/_2$ years, however, are not enough to get lasting achievements. Tanganyika is literally sprinkled with small schemes which were started, looked promising, later on stagnated because the officers changed, and are meanwhile forgotten. Not all of the officers understood the different social and cultural position of the African peasant. Some became good extension workers, others failed. The qualifications of European and African personnel varied widely. It was apparently not possible to dismiss those who were not worth their salaries. A considerable ballast of unsuitable personnel was kept on.

Justified as such criticism may be, it must also be stated that a small but well-trained staff of officers, experienced in African farming, worked with zeal and a remarkable degree of devotion, in order to promote the welfare of the peasants. The fact that Tanganyika is still economically undeveloped, even compared with other African countries, is not due to the failure of some development efforts. The explanation is rather that they started late, and with little personnel and money. A substantial proportion of what has been depicted here as British agricultural development policy was more a matter of intention and beginning rather than materialization. Compared with the small input, however, much was achieved. Agricultural officers of the British colonial administration were quite successful intermediaries of technical progress.

IV. The Development Plan for 1961/62—1963/64

In November 1960, the World Bank Report, the first extensive study on the economy of Tanganyika, was published. It contained recommendations for further economic development. They represent for the most part a continuation of the successful economic policy of the late fifties. The promotion of peasant farming without prejudice to the agricultural estates is regarded as the major task. An important supplement to this policy is the extension of the system of roads and of the educational system.

For agriculture the aims which were set are identical to those which were given priority under British administration:

- Improvement of the existing farms by providing extension personnel and small loans, organization of marketing facilities and incentives through consumer goods.
- Transformation of land use in special schemes under close supervision (settlements, plantation crops, irrigation).
- Building up the existing ranches, followed by the establishment of more ranches.
- In one respect, however, the World Bank Report differs from the current development practice. It proposes: Preparatory work on an irrigation programme, which is to begin in 1965/66 on 7,500 acres and by 1969/70 to provide irrigation for 25,000 acres annually.

The Development Plan for 1961/62 to 1963/64 followed the first three recommendations[1]. Of a total additional £ 24 million which the government wished to provide in three years, £ 5.7 million were to be spent on agriculture and £ 0.9 million on forestry and wild life. The construction of roads, for which expenditures of £ 4.2 million were planned, was largely designed to improve marketing possibilities for agricultural products. The expenditure of £ 0.229 million for "Community Development" will partly benefit the rural population.

The objective of these expenditures is to create a basis for continuous economic growth, i.e. the preparations for the subsequent "take-off". Production increases in agriculture, especially of export crops, and the increased division of labour among farmers, traders and processors are designed to:

- increase the tax revenues of Tanganyika so that the unavoidable governmental and development expenditures can be more satisfactorily covered by the state itself;
- create additional possibilities for industrial processing of agricultural products;
- increase the market possibilities for the consumer industry by strengthening the buying power of the population, i.e. of rural families.

Thus, according to this plan, industrialization is to be promoted by strengthening the agrarian basis, mainly because there are no other points at which industrial activity can be initiated. There is no dominant mining industry. The internal market for consumer goods is still too small and too

[1] The term "development plan" is incorrect. This is a development budget, which contains the additional expenditures of the various governmental organs for the three-year period. Since the additional services which the government will provide are designed to effect development, there is a distinction between this budget and the usual governmental budget. This development plan does not contain measures for development in the private sector.

splintered. It is to be expanded by extending agricultural production and agrarian exports. These measures are supported by the fact that increased agricultural exports from Tanganyika would probably have no noticeable effect on world market prices, with the exception of sisal which enjoys a favourable world market position.

The most important agricultural part of the development plan is the investment of £ 2.35 million in agricultural crops. The expansion of agricultural schools and extension work supplemented by appropriate aid in the form of equipment, grants, research, etc. is supposed to provide additional services (Table 31). The major objective of the plan is not the

Table 30. *Development Plan 1961/62—63/64*
Functional Analysis

	in mill. £	in percent
Agricultural crops	2,355	9.8
Veterinary	319	1.3
Tanganyika Agricultural Corp.	507	2.1
Water Supplies and Irrigation	2,291	9.6
Total Agriculture	5,472	22.8
Co-operative Development	265	1.1
Community Development	229	1.0
Forestry and Wild Life	887	3.6
Roads	4,167	17.5
Power	1,800	7.5
Industry (Development Corp, and Industrial Credit)	680	2.9
Education	3,270	13.8
Health	954	4.0
Urbanization & Buildings	2,875	12.0
Police & Army	2,380	9.9
Miscellaneous	951	3.9
Total	23,930	100.0

Source: Development Plan 1961/62—63/64, Table XIV, p. 14.

formation of capital in the form of capital goods — this task is to be left to subsequent stages of development — but as "human investment" in the farmers by disseminating information on technical knowledge, which is considered the ferment of economic and social change (Table 30).

There is still another important aspect to be considered. The development plan was projected into Independence and therefore into a different psychological situation. Under the British administration agricultural extension work was plagued by hesitancy and at times even by the open opposition of the farmers towards measures proposed by foreigners on behalf of a foreign government. Few farmers were able to realize that this work was done for their own benefit. The politicians regarded the enmity towards innovation of large parts of the rural population as an opportunity to gain followers, and often organized resistance against measures which were to aid the rural population in its struggle against ignorance and poverty. It was assumed by the planners that independence would bring a considerable change. NYERERE and the Tanganyika National Union intend to put an end to this politically conditioned "opposition" towards agricultural progress. TANU is to switch from opposition to a policy of support [1]. The development plan was based on the hope that political institutions will animate farmers and that experienced extension workers will have a better chance to transmit modern farming techniques. At the time the plan was drawn up, it was evident that most experienced extension workers would be Europeans; they would, however, be employees of an African government, not of a foreign power. It was assumed that political independence would establish an identity between the African people and an Afri-

Table 31. *Agricultural Projects of the Development Plan 1961/62—1963/64*

A. Agriculture (crops)		
1. *Agricultural Training*		£ 178,048
a) Agricultural Training Centre Ukiriguru	98,040	
b) Land Planning Centre Morogoro	21,401	
c) Animal Training Centre Urambo	17,625	
d) Natural Training Centre Tengeru	40,982	
2. *Extension Services*		£ 1413,845
a) Expansion of Staff and Ancillary Services	859,670	
b) Capital Improvements	224,650	
c) Fisheries	22,600	
d) Use of Draught Animals	17,640	
e) Crop Husbandry	45,000	
f) Tea Growing	57,100	
g) Plantation Crops	30,000	
h) Grazing Systems	51,000	
i) Irrigation	30,700	
j)—m) Miscellaneous	75,485	
3. *Specialist Services*		£ 179,790
4. *Research Services*		£ 222,321
5. *Subsidies*		£ 88,300
6. *Farm Institutes*		£ 116,075
7. *Agricultural College*		£ 406,950
Total		£ 2605,329
less current expenditure		250,000
Remainder		2355,329

[1] TANU became soon after independence the only political party in Tanganyika.

Table 31. (continued)

B. Veterinary Services	
1. Stock Routes and Quarantines	111,428
2. Disease Control	149,022
3. Immunization Camps	4,100
4. Buildings and Works	27,600
5. Central Laboratory	20,000
6. Tsetse Eradication	6,915
Total	319,065
C. Tanganyika Agricultural Corporation	
1. Mbarali Irrigation Scheme	£ 56,800
2.—3. Production Farms and Ranches	200,000
4.—5. Planned Settlement Schemes	250,000
Total	£ 506,800
D. Water Development and Irrigation	
1., 2., 3. Water Supplies, Surveys, Investigations and Equipment	£ 736,000
4. Water Supplies (Native Authorities)	900,000
5. Water Control	360,000
6. Water Supplies (Government Outstations and Minor Settlements)	45,000
7. Nyumba ya Mungu Dam	250,000
Total	£ 2291,000
E. Co-operative Development	
1. Staff Expansion	£ 203,190
2. Visits of Advisers	12,000
3. Co-operative College	45,000
4. Co-operative Union	5,000
A + B + C + D + E total	£ 5737,384

Note: The sums given in the Development Plan under "miscellaneous" (p. 72) for the Natural Training Centre Tenegeru and the Market study are included under 1. Training and 4. Research.

Source: Development Plan, pp. 49—73.

can government which would change the European officers' position from being an exponent of a foreign power to becoming a helper in the construction of an African economy and society. Anyhow, expansion of the agricultural training system was to make possible the gradual "Africanization" of the extension services.

Thus the priority given to agricultural extension in the development plan was based on political as well as on economic considerations. The investment made in providing technological knowledge was designed to fully utilize the receptivity to change and innovation which is usually evident during periods of rapid social change. The expansion of agricultural extension work, which according to the plan is to be almost doubled, appears to be a rather hasty measure. Experience shows that it is advisable to expand such services gradually. These misgivings were put aside in view of the favourableness of the hour.

The priority attached to extension necessarily led to the reduction of funds and personnel available for irrigation. The plan neglected irrigation proposals put forward in the World Bank Report, as well as FAO reports about irrigation possibilities. The idea was that the time for irrigation had not come yet. Preference was given to exploit the chance of receptiveness to change through extension. This chance will not last for long. It should be used as long as it is present. Irrigation may come later.

Chapter E

Agricultural Development Policy in the first Year of Independence

I. Africanization

In the first months of independence it became obvious that the realization of the plan's basic idea would be difficult. The question is whether it is wise to pursue a rapid expansion of the development services of a government at a time when its personnel is undergoing changes due to a different political situation. The attitude of rural people had been correctly assessed. Independence is echoed even in the most isolated villages. Long-cherished hopes are suddenly being fulfilled. BLOHM quotes the sayings of the Nyamwezi from Germans times:

> "One day something is going to happen, something mighty, which will overcome all Europeans of our time. No white man will be able to do anything against it. That is the belief of the oldest of the Nyamwezi." (Part II, p. 193.)

Meanwhile this event has occurred; it is welcomed in the most isolated village, is the cause of great material hopes, is considered the symbol of newly-won human equality and has produced an extraordinary willingness to accept social change and innovations, to put aside tribal customs, to accept Swahili, a language which is not the mother tongue of most of the people, and even to accept unusual administrative measures. Surprisingly enough, there is no evidence of hostility towards Europeans, at least not in the villages. All over the country European officers are finding that more than ever before they are regarded as helpers. They no longer need to look for receptive farmers. From many places it is reported that for the first time in the history of the Agricultural District Office, farmers, headmen and TANU secretaries are requesting advice in agricultural matters. This is indeed an opportune time for introducing innovations.

Yet, the positive circumstance of the hour cannot be fully exploited. Most of the European officers who know the language and country have left. Irreplaceable capital in the form of experience and knowledge has gone with them. This capital could never have been more effective than in the years immediately following independence. Qualified African extension workers are so few in number that even with the best of intentions they are unable to fill the gap. It was assumed that an African government could work with a largely British agricultural administration which is gradually being Africanized. This assumption did not come true. The theory of a "multiracial society" in which the individual is judged on the basis of his qualifications regardless of the colour of his skin, is equally unrealistic. Africans desire to gain positions, repute, and control in every field. This aim is so prominent that economic considerations "hardly weigh as much as a feather". Tanganyika's administrative staff is being Africanized more rapidly than that of many other countries on the continent, even though there are fewer trained people here than elsewhere.

If the British officials had remained in responsible positions and in great numbers, the higher and middle TANU functionaries would have become discontented. Africanization acted as a safety valve in the first years of independence. It was not confined only to the police force, the administration, radio etc., i.e. to the positions of power, but was applied as a general principle also to technical services. By the end of 1962, almost all higher agricultural positions on a district and regional level were held by Africans. This is a rapid change for which the Agricultural Department was not prepared. Africanization of staff strted in the 1940's and included in 1955/56 hardly a dozen positions. It did not get into its stride before 1959. In mid-year 1962, there were 65 Africans employed by the Extension Division whose complete staff numbered 162 (including those on leave and in training). By the end of 1962 these 65 held almost all of the higher positions. Those Europeans who wish to remain are systematically placed in research and teaching.

Africanization did not mean the expulsion of European officers. As a rule the African filled a vacancy which had been created because the British predecessor was no longer satisfied with his working conditions. Approximately two-third of the Agricultural Officers employed in 1961 left their positions in 1962, or were thinking of doing so. This is more than had been expected in 1961. At that time it was assumed that about one-third would leave the country after independence. It was obvious that no more careers could be made in the colonial service. For the younger, well-qualified officer to leave was a necessity. It was expected that those who could not psychologically make the sudden change from colonial times to independence would leave the service.

However, the exodus included also those who would have remained and performed valuable service. Financial incentives to remain have indeed been made. The British government contributes a considerable extra allowance to the wages paid by the government of Tanganyika. Upon termination officers receive substantial compensation payments. There was the chance to keep officers between the ages of 40 and 55 until the time when they would like to be pensioned. Most of them went away. Some even had the feeling that their departure was welcomed. Perhaps it is asking too much to expect trust between former colonial officers and an African government in the first months of independence. The damage to the country, however, is obvious. As a result of the rapid Africanization, many persons have been given much responsibility too suddenly. Work under such superiors is usually not very gratifying. Work is accumulating for those remaining. They have to take over what others have left, in addition to their own tasks. The government's desire to induce activity at every level means that the Agricultural District Officers are flooded with paper work and have to take on responsibilities which lie outside their fields of interest and experience. There is little time left for actual extension work which motivated their entry into the government service. They can no longer do what as technocrats they consider useful. All these factors have had a demoralizing effect among the European staff. In view of such conditions it is obvious that particularly the capable ones go. Africanization has been rather arbitrary. The lucky one is not always the best qualified man. Apparently an academic diploma and seniority in the service are decisive qualifications for promotion. Experienced and talented individuals who do not have professional or academic training are easily forgotten. Promotions are made from the top down. Young college graduates with little professional knowledge are given positions which normally require 20—30 years of experience. Positions are given to those candidates who are available when vacancies occur. Since independence, there has been a campaign to find suitable scholarship applicants. When these students return after several years of thorough training, they will find that the very positions for which they are qualified have been filled by less qualified "stay-at-homes". For the present at least, little account is being taken of a measure essential for successful rapid change in personnel: the rapid release of those who prove unsuitable. This is even more true of the new administration than it was of its British predecessor.

The Agricultural Department includes a number of African officers who have long experience in the profession. Some have been most successful extension workers. Yet they are too few in number. The placing of Africans in the technical and administrative services in Tanganyika began very late. This was one of the greatest omissions of the British. Most of those who were rapidly promoted had just begun to practise the profession. They had become used to clear, specific technical tasks. Experience indicates that it is easier to learn the techniques of coffee or cotton cultivation than to make economic judgements. The fact that time, the means of transport, soil, water etc. are economic values which can be measured in money, and that unsuitable workers should be released as soon as possible in the interest of development, are strange conceptions. If, as had been expected, a framework of experienced European officers had remained, the new and inexperienced personnel could have been effectively placed and trained. This opportunity was lost.

Such difficulties are normal in times of political change. In face of the late educational preparation for transmitting responsibility to Africans, it is noteworthy that the services did not break down completely. Yet it is questionable whether the conception of the Development Plan is still appropriate. The Report of the World Bank, referring to the situation in 1960, states:

> "Fortunately Tanganyika has an efficient administration. This report can assume that correlated programmes will be carried out systematically and correctly. If this assumption were not correct, the report would have been drawn up differently and future developments would have been assessed with more reservation." (p. 4.)

Meanwhile conditions have changed. The new administration will have achieved a great deal if it is able to pass through the current changes without too many breakdowns. The administration will need some years to consolidate and to sift out qualified personnel.

II. African Socialism

In addition to changes in personnel there were those in agricultural policy. British agricultural officers are pragmatic technocrats, usually with middle-class attitudes and an unmistakable tendency towards economic planning. They spread technical innovations and thus indirectly promoted social and cultural change. In contrast, the new government wants to create through direct political and social changes an economic receptivity which will lead the farmers to seek innovations on their own initiative [1].

---

[1] KAMBONA, The Minister of the Interior has stated: We want to awaken those who slept during colonial rule.

Consequently, the African government cannot simply continue to administer the country pragmatically. In order to live as a human being man needs an idea, which is in front of him and above him, which gives him guidance, which helps him to select and which might even compel him to choose certain approaches [1]. The efforts towards this lead to the formulation of an "African socialism", for — as NYERERE writes: "no economically underdeveloped land can afford to be anything other than socialistic".

There is so far no clear definition of African socialism. The following quotation from SEKOU TOURÉ concerning Guinea, which is largely applicable to Tanganyika, gives some idea of the basic thinking behind African socialism [2]:

> "The rest of the world assumes that we will adopt this or that political programme and adapt our political activities and our realities to that programme. We have a completely different conception. We start with the realities and the objectives, and from these we develop a programme of action which we will implement by using any available accelerative means." (Spearhead VI.I, No. 8.)

Thus, African socialism is pragmatic. It contains elements of the socialism of both Western and Eastern Europe, especially of the Fabian type. It has roots in African tribal tradition. NYERERE places much significance on overcoming the colonial de-Africanization of minds by returning to the cultural and moral African heritage. The individual was given security by the tribe which demanded from him in return co-operation and tribal turnouts. Sometimes fields were cleared communally. Principles valid for the tribe are to be transferred to the state, which is to be considered as an extension of family and tribe. Tribal traditions transmit a classless society. Tribe members are equals. Tribal traditions discourage individual, differentiated activities which would endanger the unity of the group and thus survival under conditions prevailing in former centuries. African socialism takes advantage of these traditions. The formation of a middle-class is explicitly declared undesirable. Not the ambitious individual is to be promoted, but the group. The peasant shall not strive in isolation to improve his life, but in communal action. Tribal traditions say that land is owned by the group, not the individual. Consequently, African socialism is replacing the tribe by the state or the co-operative and these are to decide on the allocation of land capital. However, pragmatism can go far. Partnership between a private firm which has a share in the business and acts as

[1] ANTWEILER, A.: Entwicklungshilfe, Versuch einer Theorie. Trier 1962.

[2] See also: SENGHOR, L. S.: Les Données du Problème. Conference at Dakar, December 3rd to 8th, 1962. — BIOBAKU, S. O.: Notes for the Report of the Nigerian Delegation, ditto. — KANOUTE: Socialisme africain, expression de l'humanisme africain. In Afrique Nouvelle, Dakar 1962, Nr. 799. — MBOYA, T.: African Socialism. In Transition, Kampala, March 1963. Nr. 8.

manager, and the capital of an African co-operative or the state can certainly be considered an institution which is acceptable in African socialism.

The economic policy of Tanganyika is, for the time being, largely synonymous with agricultural policy. Thus, agriculture becomes the testing ground of African socialism. In a period of rapid change it is understandable that many aims, both realistic and unrealistic ones, are being considered:

— We must work hard. Land is there. Everyone should take a hoe and begin to cultivate.
— We must throw away the hoe and use tractors.
— The time of the peasant is over! The future belongs to mechanized large farms.
— Peasants should organize co-operative land use.
— The TANU-Youth-League shall be the carrier of innovations in the villages.
— Introduction of a compulsory National Service for young men.

A separation of appropriate measures from inappropriate ones will probably take place rather quickly, due to the shortage of money and personnel. Also, there is the awareness that hasty measures give rise to opposition. "They (the farmers) change slowly", says Parliamentary Secretary Barongo — "they tend to oppose every measure which they feel is being forced on them". In this respect the new government is very careful. It abstains — in an almost exaggerated manner — from endangering the existing identity between the people and the government.

However, there are clear intentions of a "levée en masse". The whole population is to participate in economic development. "Independence and work" (Uhuru na Kazi) is the slogan. Possibilities to this effect are in peasant agriculture rather than in industry or plantations. By providing the political impetus, Nyerere would like to induce the "Baba Kabwela", the poor peasant, to become more active, to develop wider interests, to participate in communal action, and thus to help himself. The emphasis is not on active individuals, favourable areas, or willing tribes, but rather on the uniform progress of all the people who inhabit this huge, thinly populated block of land called Tanganyika.

III. "The People's Plan"

1. Institutions

One of the first measures taken by the independent government was to bring politics into the administration. TANU-politicians, not African administrators, took the place of the British officers as representatives of the government. Politicians have administrative officers as secretaries. The "Regional Commissioner" is the highest official of the region, the "Area Commissioner" the highest official of the district.

Immediately after taking office, Regional and Area Commissioners were entrusted with development tasks. In accordance with the idea of a "levée en masse", the three-year Development Plan, which is still the standard but which affects only the budgets of government departments, was supplemented by a "People's Plan". This is not a plan in the usual sense, but one in which "the people" are to set local targets and the Regional and Area Commissioners to see that they are reached, that there is no absence of enthusiasm and that the enthusiasm for independence is led into productive channels. The aim is:

— to increase agricultural production. Each village and each district is to set its own targets. Regional and District Commissioners and their staffs are to give encouragement, to coordinate and to see that people make the greatest efforts to carry out their own plans in order to ensure that these plans are fulfilled or exceeded. (Communication of the Minister of Local Government of February 27, 1962.)
— to relieve the pressure on the Treasury by inducing the population to help themselves, i.e. to perform unpaid communal work in the construction of roads, dams, better houses, erosion control, wells, local water supplies, etc.

The following institutions are to help in the achievement of these objectives:

— The *Regional Development Committee* is made up of various officials, TANU secretaries, members of Parliament and representatives of the private sector. It is to give guidance, co-ordinate, transmit technical aid, examine and edit district plans, and place the regional plan before the central government.
— The members of the *District Development Committee* correspond to those of the Regional Development Committee. As occasion arises, representatives of co-operatives or of economically active missions are added. The District Committee works out a general plan for its area, gives suggestions as to the crops which peasants should plant, co-ordinates and concerns itself with local taxes for financing the plan. The district plans are examined by the Region.
— The newly organized *Village Development Committee* is expected to become the most important of these institutions. This local-action committee is to be composed of leading personalities of the village [1]. (Headman, TANU Secretary, Chairman of the Co-operatives, leading farmers, the teacher, Agricultural Field Assistant, etc.). Village commit-

[1] It is difficult to determine to what extent such committees are appointed or elected. In Tanganyika it is the custom for participants to sit under a tree and to discuss until they have reached agreement (J. K. Nyerere, in: Transition, Vol. I, No. 2, p. 9).

tees are expected to work out their own local development plan. The idea is that no outside office should set the local targets, but rather the people themselves. They make their own decisions concerning action to be taken, the amount of new land to be farmed by each individual, which roads to build and which to improve, how much unpaid work each individual is expected to perform, etc. The village plans are sent to the district office for examination and approval, so that losses in time and equipment are avoided as far as possible.

The vertical division of responsibilities is supplemented by a horizontal one:

— The *general administration* and the responsible Regional and District Development Committees are to give guidance to the work at village level, to supervise, to organize technical aid and to supply equipment.
— Staff members of the *Department for Community Development* are to take over the role of specially trained animators and co-ordinators. They are responsible for inducing the grouping of people, co-ordinating between village and district, aiding in the choice of realistic projects and stimulating the wants of the population by providing adult education, campaigning against illiteracy, and instructing the women in respect of clothing, child-care, nutrition and health.
— *TANU,* and especially the TANU Youth League, is to provide political stimulation, i.e. to organize local public opinion and to strengthen the influence which progressive neighbours can exert on apathetic ones. Occasionally TANU is termed the "whip" of village progress.
— The *Agricultural Department,* which until now has provided most of the stimuli in the villages, is placed in a different position. It no longer acts as instigator of initiative, but rather as an adviser. Agricultural officers are no longer to look for interested peasants, but rather the Village Development Committees, Community Development workers, and TANU Youth League members are to see that the peasants consult the agricultural officer on their own.

2. Some Beginnings and Achievements

In the summer of 1962, the work on the "People's Plan" began to take shape. A careful assessment can be made only after some time has passed. This study, which covers the period ending early in 1963, can take into account only the initial stages, the early achievements and the first indications of difficulties.

a) Capital Formation through Unpaid Labour

The old African tradition of "tribal turnouts", which under colonial rule was hated as forced labour, has been successfully rejuvenated and is known as "self-help". It has become a public event or even a social occasion. Participation is not as binding as the clearing and farming of additional land, and offers the pleasure of "seeing the wealthy ones sweat". This pleasure is provided by Indian traders and their wives who are not used to physical labour. For this reason alone villagers oppose the fulfilment of communal responsibilities with monetary or material contributions.

Following is a record of projects completed in 1962 [1]:

166 schools
134 community houses
31 hospitals
66 dispensaries
10,000 miles of feeder roads
6 bridges
10 drifts
903 miles of road improvement
321 miles of water pipes
313 wells
142 small dams for water supplies
110 miles of furrows
10 watering places

In addition, there are co-operative centres, housing units, latrines, football fields, etc. The value of these investments is estimated at £ 200,000. The government contributed £ 40,000. Of course, these figures have to be interpreted with care. The desire to show accomplishments certainly leads to a tendency to overestimate.

Nevertheless, the number and popularity of self-help-schemes exceeds government's expectations. It is said that in 1962 one million people participated. Supply of technical guidance and material is not always able to keep up with local demands. For the newly-appointed Area Commissioners the self-help projects have become the yard-stick of their work.

b) Cultivation Targets

The establishment of Village Development Committees began in the summer of 1962. In June 1963 they existed almost everywhere in the country. Without going into the question of how good they are — this is an impressive organizational achievement.

The first task of the Village Development Committees was to set cultivation targets. Agricultural Officers saw in district and village plans a chance to put through improvements which had been attempted with so

[1] Achievements of the "self-help schemes" as published in the Newsweek of December 24, 1962 and Tanganyika Standard up to March 1963.

little success in the past. They recommended targets for terracing, tie-ridging, pulling-up of old coffee trees and their replacement by new ones, reduction in the number of cattle in overgrazed areas, popularization of better seeds, propagation of better husbandry and other measures to increase yields per acre. These recommendations were for the most part not adopted. They did not fit into the concept. Simple and popular targets are necessary to arouse and maintain the enthusiasm of the masses for self-help. Erosion control, the pulling-up of old coffee trees, and especially the reduction of cattle, are extremely unpopular measures. Efforts made to implement such measures have led the peasants in some places to regard the Agricultural Department as nothing but an institution "to annoy and upset the farmer". Therefore, the activation of the masses is to begin with projects which they themselves determine in the Village Council. The principle of increased yields per acre is for almost all of the peasants who think in static terms — and even for the majority of TANU secretaries — a strange and unfamiliar conception. In their view, production increases are identical with the expansion of the farming area. Concrete targets can thus be set. A plan which calls for 50 per cent of the farmers to husband the cotton correctly is one which can hardly be measured and politically exploited. A target calling for each farmer to plant an additional 1 or 2 acres is much simpler to propagate and perhaps even to control.

Therefore, Village Committees all over Tanganyika are setting targets for expanded cultivation. They are often Utopian, calling for each farmer to plant 10 or 15 acres, or they exceed the total acreage of the village. The Agricultural Officer in the district office usually alters the plans. The resulting targets are of the following types:

Coastal area near Tanga:

Every family is to plant an additional 2.5 acres, of which 0.5 acres are to be planted with cashew trees and 0.5 acres with coconut palms.

near Tabora:

The additional cultivation of 2 acres of cash crops and 1 acre of subsistence crops per family.

near Morogoro:

The additional cultivation of 1 acre of cash crops and 1 acre of subsistence crops, the planting of 50 cashew trees and 50 coffee trees.

Targets for 1962/63 were frequently thought out and put on paper in the Agricultural District Office without the villagers' knowledge. In 1962, everything had to happen in a hurry. Nevertheless, it is apparent in many areas that more than the usual has been cultivated. This is certainly worth noting.

c) Co-operative Land Use

Parallel to cultivation targets, sometimes directly connected with them, efforts are made to introduce co-operative land use. There are several approaches:

Communal Farms. Frequently villagers clear some land communally in order to fulfil their targets and to obtain money for community purposes. Communal land clearing belongs to the traditional way of life Nowadays these clearings are called Communal Farms. They are usually planted communally with one cash crop. In addition, participants cultivate their own fields — mainly with subsistence crops. The organization of Communal Farms changes. The cleared land is either

— divided among the participants — everybody being responsible for weeding and harvesting one plot in the area,
or

— part of the area is divided. The rest is farmed communally and the income from it used for community purposes,
or

— all land of the cleared area is cultivated, weeded and harvested communally. The yield is divided among the participants or goes to some communal purpose.

The number of Communal Farms early in 1963 was estimated to exceed 600. They vary in size between 1 and 15 acres. In some locations, participants employ wage labour.

Farming Associations: Another form of co-operative production has developed, thanks to a certain approach in loan policy. Farmers who form a Farming Association — a non-registered co-operative which incorporates at least 20 members — can apply for loans on favourable terms. However, they have to use the money for productive investments in a co-operative enterprise. Applications usually concern the purchase of a tractor. There are two different approaches to this:

— Co-operation is limited to the use of the tractor. The group tractor cultivates the fields of individual farmers. They may be participants or not. Tractor services have to be paid for.

— Co-operative tractor use is coupled with co-operative land use. Here is one example: 100 farmers have, in addition to their own fields, cleared 200 acres of land and divided these among themselves in equal plots. Half of the land is planted with maize, half with cotton. Neighbouring plantations and traders were induced to contribute money for the establishment of a tractor station, which is managed by a secretary. Tractors do the ploughing. Participants are responsible for planting, weeding, and harvesting. An Assistant Field Officer is appointed as agricultural adviser. Tractor costs are to be covered by crop returns. There is no clear organizational set-up. The first year was a failure.

In other cases Farming Associations concern the co-operative cultivation of plantation crops. Here is one example with sisal:

— 168 farmers who own individual fields, cattle and coconut palms, clear communally 320 acres of bush and begin to plant sisal. Each participant receives a card recording the number of working days. This card is to serve as the basis on which the later crop is to be divided. The area is farmed communally. On the basis of the initial efforts the government granted a loan of £ 2,300 for additional clearing, rooting and ploughing by a contractor. Maize from the USA was distributed free of charge. There is no clear organizational set-up as to weeding, cutting and selling. The participants simply got together and started work. One year after the beginning, the area was neglected and heavily weeded.

Co-operative Settlements. Substantial hopes are placed on co-operative settlements. The labour policy of the new government is aiming at well-paid labourers in towns and estates who are working a regular 8 hours a day and who have their permanent home close to their place of work. Inherited from the past are low productivity, low wages, unreliability in daily appearance, and employment of migratory workers [1]. In order to achieve its objective, the government allowed the wages to be doubled within two years. The necessary consequence is a rapidly increasing number of unemployed [2].

— There are theoretical considerations which indicate that the government's labour policy is sensible in spite of its causing unemployment. There is — or at least was — substantial scope for rationalization in the use of labour in estates and towns. A significant increase in wages could be absorbed without increasing substantially wage costs per unit of output. Thus, a certain jump in wages was not so much a burden for the employers as for those employees who had to be dismissed. There is, however, in particular close to larger towns such as Dar-es-Salaam and Tanga, ample arable land available. Those who have become unemployed will have to go back to the land sooner or later. There is

[1] In 1958 Guillebaud estimated that the work done by a farm worker in Europe was approximately equal to that of 5 workers on sisal farms in Tanganyika. The performance of sisal workers in Java is said to have been two to three times as high as in Tanganyika.

[2] As to Dar-es-Salaam, in 1962 the number of unemployed was estimated at 25,000 and that of half unemployed at 12,000. The number of Africans employed in agriculture fell from 220,199 in July of 1959 to 179,400 in July of 1961.

no other outlet for them but to take the hoe and to cultivate. This, in due course, will not only lead to additional subsistence farming but also to some additional supply of cash crops. Consequently, a certain increase in wages does not reduce output in plantations and towns. It increases peasant production, however.
Furthermore, the substitution of large numbers of poorly paid workers by a smaller number of well-paid ones is having effects on tax revenues and import duties. Firstly, there is some revenue forthcoming from additional production by those who had to go back to the land and to produce cash crops. Secondly, higher paid workers will use additional income with a preference for items which are highly taxed: beer from breweries, bicycles, radios, etc. All in all, under the condition of free availability of usable land — which is true for the coastal areas in Tanganyika — a certain jump in wages can effect increases in the gross national product and in the revenues of the Treasury.

The unemployed, however, are discontented. The safety valve for them has to be the return to the land through participation in co-operative settlements. There is no need for anybody to be unemployed. The Minister of Labour, Kamaliza, does not hesitate to say that the unemployed are lazy. They should take the hoe and cultivate land; there is enough of it.

The significant difference between settlements and Communal Farms or Farming Associations is that, in the former, participants live entirely in the scheme area and that they farm nowhere else except in the settlement. Several types of settlement can be distinguished:

— spontaneous settlements
— planned government schemes for urban unemployed
— TANU Youth League Farms and National Service Farms.

Spontaneous Settlements. Discontent with the employment situation in towns and plantations has given rise to the formation of spontaneous groups which clear and cultivate land communally. Here is one example:

— A group of 50 cleared 100 acres of land and planted cashew nuts and manioc. The choice of plot was made on the advice of the Agricultural Officer. The settlement received support in the form of seeds and US maize. In the first months, half of the participants left the settlement. The remaining ones showed good spirit. A good manioc harvest was the first success. Each participant received 500 shs in cash. The aim now is to buy a tractor. The leader of the group is a man of very strong personality.

Planned Government Settlements. In order to be able to provide definite agricultural working places for the unemployed, several more or less planned settlements have been established. He who demands work is asked to go there. The Agricultural Department chooses suitable land, distributes seeds and tools free of charge. The administration delivers maize and milk-powder provided by the US until the first crop is harvested, i.e. aid valued at about £ 90 per family. Where necessary, the government takes over the provision of drinking water. Each participant can be granted a loan up to 300 shs.

Support is given only to those projects — at least this is the intention — which are approved by the Agricultural Department and include at least 20 families who make their homes on the settlement and are considered by the administration and TANU to be reliable persons. In October 1962, 13 such settlements existed. They are among the schemes which are especially promoted by TANU. There are far-reaching conceptions of model village settlements, equipped with a school, a dispensary, a tractor station, and an organizational set-up which corresponds to that of the Moshav System of Israel. Until now, the realization of these settlement plans has suffered because very few unemployed are willing to come.

— **An Example.** Not far from a town which has several hundred unemployed there are 30,000 acres of moderately fertile land available. The annual rainfall of 27 inches is rather low, but is considered relatively reliable. The plan runs for 400 settlers. Each of them is to receive 1 acre of irrigated land for rice and 6 acres of dry land for cotton, groundnuts and manioc. The participants receive tools, food, and plants. Under the direction of an Assistant Agricultural Officer they are to clear the land themselves and to construct roads and irrigation channels. The granting of a motor pump is being considered. It is hoped that the organization "Freedom from Hunger" will provide food in the first year. The use of land, tractors, and water, as well as sales will be organised by a settlers' committee. However, not enough people are taking part. There were not more than 35 applicants when the scheme started. Potentially interested persons are discouraged by the obligation to live on the scheme. Experience indicates that it is easier to persuade peasants to participate and to remain than to attract unemployed people from the town.

There are other planned settlements apart from those for the unemployed. Settlements for the Watussi who have fled from Ruanda-Urundi to the neighbouring areas of Tanganyika are similarly organized, though the purpose is different. Kikuyu from over-populated areas of Kenya have been settled near the last station on the Mpanda Railroad in the Central Region. The long-range objective is to make the railroad economically sound by supplying more goods.

TANU Youth League and National Service Farms. With approaching independence, the TANU Youth League took up farming. One reason is the desire to serve the nation through additional agricultural production; another, the idea that modern farming can take place only on large farms;

a third, the wish to engage in group action and co-operative enterprise. The TANU Youth League appealed to youngsters. They joined a kind of voluntary labour service. In some cases land was cleared. The enthusiasm was so great that they did not even think of asking for guidance and advice. It was believed that hard work — which has indeed been done — and enthusiasm for co-operative action is enough.

In due time the approach changed somewhat. Discussion early in 1963 centered around the establishment of a National Service for youngsters. They would do agricultural work in 1200 acre farms, equipped with tractors. The youngsters would be organized by the TANU Youth League. Agricultural Field Assistants would provide the technical know-how. The first National Service Farm was established in the Central Region in 1963.

Villagization. More importance is attached to "villagization" than to any of the above-mentioned settlement types. There are few villages in rural Tanganyika. Peasants usually live in huts on their land. Settled countrysides are sprinkled with isolated huts or hamlets. President NYERERE proposes to concentrate the population in villages. This is expected to bring the following advantages:

— Land can be consolidated in fields large enough for the employment of tractors.
— Obligatory land use can be organized — comparable to medieval practices in middle Europe —, i.e. every farmer has to cultivate the same crop in a certain area. This may further the introduction of rotations, improved seeds, mineral fertilizer, erosion control works etc. It will be easier to prescribe planting time, the acreage to be planted and the quality of husbandry. It will be simpler to apply controls.
— The traditional pattern of settlement makes it financially impossible to provide farm households with healthy water. In villages it will be much cheaper. Furthermore, people will have better access to schools, dispensaries, markets etc.

Villagization is to be started with 5 to 10 trial schemes. They will be models. Establishment is not planned in already settled areas but on new land. Schemes are planned to include 250 families, whereby every family is to have a home plot of 1 acre. Another 9 acres with cash crops are to be farmed co-operatively.

d) Co-operative Trade

Co-operative trade is pushed even more than co-operative production. There is a double purpose behind it. Firstly — and this is the most important aspect — co-operative trade will lead to Africanization. It is an indirect means of replacing Indian traders by indigenous activity. Secondly, co-operative trade is an important approach to replace private traders and thus towards African socialism. Under the British administration, local co-operative efforts were supported and channelled. Where people had no interest, no specific incentive was given. To the new government this is too slow an approach. Their officials are to proceed more directly. Hesitancy is regarded as requiring a decolonization of minds. People are not forced to trade co-operatively. There is, however, systematic propaganda in favour of co-operatives, and there is also some pressure. As soon as feasible — and perhaps even sooner than that — co-operatives are given an exclusive trading right in certain commodities. Indian traders are to be squeezed out. They are expected to become employees of the new co-operatives, thus helping them with their great experience in trade.

Due to the new efforts, the number of registered co-operatives increased from 760 to more than 1000 in the first year of independence. It can be observed in some places that the peasants bring their produce to the co-operative although Indian traders pay higher prices. There is undoubtedly some degree of willingness on the part of the rural population to make sacrifices in order to establish co-operatives.

Co-operative development, however, will conform to government market policies. Rural co-operatives are not expected to export directly. Price and export policy will be the concern of Marketing Boards. Agricultural products will pass from *communal or co-operative production to co-operative trade* and from there to *state-run marketing boards* which organize exports and internal markets.

e) Community Development

Institutional changes "from above" require explanation "below". There is the conviction in the new Tanganyika that the peasants will co-operate with enthusiasm if every change is explained thoroughly and with patience. The new government clearly shies away from outright compulsion. The idea is that changes which are desired shall be explained so thoroughly that they become in due time the outright wish of local people.

It is up to staff members of the Department for Community Development to explain, while TANU is much more considered as an agency to provide the necessary push. Community workers are to prepare the social changes and to iron out misunderstandings. Explicitly, they have to see what people want and to try to get done what is wanted — thus creating an atmosphere of trust. This trust again facilitates the introduction of innovations which come from above.

Community Development work was just gathering momentum early in 1963. So far, it mainly consists of adult village schools teaching reading and writing, running community centres, instructing women in hygiene, nutrition and children's care. In mid-1962 it was reported that 156,000 adults learned to read and to write and that 30,000 women participated in lectures about home economics.

3. Some Difficulties

Independence and efforts to implement the "peoples plan" have brought unusual activity into rural Tanganyika. One must not ignore, however, the difficulties which develop.

a) Shortage of Personnel

The obstacle to agricultural development under these changed circumstances is not so much lack of money, of technical know-how or of interest by the peasants, but lack of personnel.

To channel the willingness which exists so as to turn it into effective production requires clear guidance from top functionaries, careful and judicious local leadership, and co-ordination among different offices. This is obviously not easy to attain. The framework of British officials is no longer intact. Large numbers of inexperienced personnel are employed and insufficiently supervised. Salaries are out of proportion to productivity. They are much higher than is justified in an underdeveloped economy [1]. Desk work has increased considerably at the cost of extension work.

African administrators — trained by the British and influenced by their technocratic attitudes — do not always approve of the working methods of their new superiors, the TANU Regional and Area Commissioners. The latter, on the other hand, do not always trust those who faithfully served the British. In some places there is a tension between TANU officials and traditional authorities, i.e. the so-called "hidden leaders". There are cases where peasants clinging to their tribal language consider the Swahili-speaking government as a continuation of the rule of foreign overlords. Tribal feelings are strong. The sudden dismissal of chiefs from public functions, which outwardly seemed to have caused no difficulties whatsoever, is much talked over and regarded with concern amongst traditionally minded groups.

In some places Village and District Development Committees are known as "talking councils". Their members are not always those who want to serve the nation. The councils include the ambitious ones or those who "like to talk". Unrealistic targets are set, unreasonable projects chosen. Roads are constructed where they are not needed. Self-help schemes are begun which cannot be accomplished locally. The government is thus compelled to put money and personnel into schemes which do not have

---

[1] Colonial rule left a most dangerous heritage as to salaries. Government officials are paid according to what was paid to British personnel. There have been reductions, but certainly not enough. Thus college graduates receive starting salaries, which are approx. three times higher than in India and two times higher than in Japan, a country of considerable wealth nowadays. Sooner or later Tanganyika will be compelled to reduce salaries drastically if internal money is to be made available for economic development.

priority. "Take the hoe and cultivate this land, the government will help those who help themselves" is said before the land has been tested for its suitability. New offices are established in the most expensive street of a town. The personnel drive around in expensive Landrovers where a bicycle, a motor cycle or cheap robust cars — like the French "deux-chevaux" — would do.

Economic use of officials and technical personnel seems to be particularly difficult. The services of expensive specialists are asked for without realizing the cost. It is not always clear to those deciding such matters that is does not pay to attach an agronomist, trained for years overseas, to a 100-acre scheme. The remaining British officials who try to straighten things out find interested superiors, but also others who think or say that they are obstructing.

There is a danger that the willingness to change may be wasted, because the institutions and their personnel cannot master the situation. There are in fact few countries in the world where the vastness, the lack of communications, of education and discipline provide as many handicaps to the desired "levée en masse" as in Tanganyika.

b) The Economics of Self-Help Schemes

As to self-help, the problem lies in the choice of schemes. Those which are wanted locally are chosen. This, of course, explains their popularity. In demand are schools, dispensaries, feeder roads. These investments however require upkeep and personnel. The central goverment is thus facing additional obligations which do not correspond to national priorities.

There is a need for a change in self-help work towards more productive investments: upkeep of roads, erosion control, reforestation, etc. Most important are small irrigation works. These, however, create a maze of intricate problems. Self-help is based on the principle that everybody is taking part, but only those with land in the irrigation area reap the benefit. Consequently it will be neccessary to register input and charge the beneficiaries accordingly. Thus we are again facing a task which is likely to make self-help unpopular, or even impossible without direct compulsion.

Another problem is the question of participation. Up to now everybody is expected to participate: officials, traders and professional men. Masses of people are thronged together so that work is hardly possible. Productivity per man is negligible. And for professional people, officials and functionaries, this is a waste of most valuable time. Sooner or later self-help will have to be organized in such a way that obligations towards the community can be rendered either through work or through money. Escaping sweat through the payment of money, however, is extremely unpopular. People want to see the trader sweat.

c) Limits of Cultivation Targets

Cultivation targets are reasonable in areas where land is available and left uncultivated because of apathy. Targets are unreasonable for areas in which there is a shortage of land, for instance on the slopes of Mt. Kilimanjaro and Mt. Meru. As a rule, however, there are considerable areas of marginal land which could be farmed. The peasants pay no wages. They are the only ones who can make use of marginal land. Problematic is the question whether it is really apathy which limits cultivation. Planting time brings a peak in the demand for family labour. Cultivation of additional acreage is accompanied by the danger of neglected weeding. It is quite conceivable that an increase in acreages will reduce total output because of lower yields per acre.

Another question is whether it is wise to set targets which are outside control. The country is wide, fields are small and scattered over larger areas. Furthermore, checking the acreage of each household is not a sufficient answer once weeding is neglected. It is certainly impossible to inquire into the quality of plant husbandry. It may be practical to set obligatory targets for minimum cultivation of manioc as a famine reserve. More can hardly be done by way of control and enforcement. Thus the setting of cultivation targets means little else but an appeal to the peasants to feel obliged to act as the group does, to conform by planting more. This appeal is widely echoed and is thus a reasonable measure. It remains open, however, as to how long it will suffice to induce additional efforts. Moreover, one has to consider that putting more land into cultivation is not the answer to Tanganyika's production problem. The answer lies much more in higher yields per acre.

d) Problems of Co-operative Land Use

Communal land clearing and planting is part of the tribal heritage in many parts of Tanganyika. The peasants work together, sing together and drink their beer together in the evening. This makes heavy work easier. Weeding and harvesting, however, is an individual affair. The question today is whether communal and co-operative production can cross the customary borderline between individual and group effort.

A sensible approach to get more production without risking government money are *Communal Farms,* as long as the peasants are interested and co-operating. Clearing consolidated stretches of land through communal effort makes it easier to introduce innovations. It can be expected that all participants will conform in the use of seeds, insecticides or mineral fertilizers. If trees and roots are cleared, the use of a tractor becomes feasible where high price cash crops are grown. Dividing the cleared area in plots which are under individual care is more promising than communal farming.

The latter approach is, however, worth trying. Little damage is done if it fails. The government has lost no money. The subsistence of participants is secured through production from individual fields which are situated outside the Communal Farm.

Farming Associations deserve more care because government loans are involved. A failure means the loss of capital which might have fostered development elsewhere. It is probably necessary to promote the use of tractors and certainly unavoidable to subsidize it. An approach, however, combining (1) co-operative tractor use with (2) co-operative land use on (3) marginal land is of doubtful value. Participants usually farm their own fields. The co-operative enterprise is something additional. As everywhere in the world, efforts will be concentrated on the individual field. There is already a tendency among Farming Associations to employ seasonal wage labour. Lack of a clear-cut organization is evident. It must be assumed that a substantial proportion of loans will not be repaid.

The situation is different with settlements for the unemployed, as long as they cost no more than testing the land and providing seeds, hand tools and food for one year — supplied as a gift by the United States of America. This procedure is cheap enough to be considered acceptable. *Spontaneous settlements* frequently correspond to this condition. They are an appropriate approach.

Planned government settlements, however, are costlier. They are of political significance. They receive high ranking visitors. Land clearance is hard work. The settlers cite the speeches of politicians. It is said that the government will help those who help themselves. The settlers show more or less significant initial achievements and expect substantial help: tractors, a supply of drinking water, a community centre, and loans in order to hire contractors for clearing. It is hard to turn down their wishes. Gifts are given, and loans provided. This in turn is corroding the initiative for self-help. What has been granted to one settlement is required by another. Settlements start to become expensive. The high input of administrative services are hardly considered by anybody. Land, however, is usually marginal. Crops can be grown, but the yields do not justify bigger investments in planning. It is difficult to obtain participants. Everybody who comes is taken. The settlers are expected to run the enterprise co-operatively. There is no "production under close supervision". The chances are poor for the loans to be repaid because there is no "one-channel marketing", proposed or enforcible. All in all, planned government settlements tend to acquire all those characteristics which normally lead to failure. The situation on TANU Youth League Farms is similar.

It is probably wise to have some planned settlements for reasons of political expediency. The unemployed who ask for work can thus be offered a definite place to go to. As long as there are some schemes, the unemployed cannot claim that the government does not take care of them. They are however no answer to the production problem. The input-output relationship is likely to be too unfavourable.

The idea that an agricultural labour service will supply cheap labour for *National Service Farms* is probably too optimistic. National Labour Service in a low wage country is usually more expensive than wage labour. It is in any case difficult to run large mechanized farms on marginal lands — as are abundantly available in Tanganyika — without incurring great losses. It will be even more difficult within the framework of National Service Farms. The Agricultural Department can provide extension workers and agricultural administrators but hardly farm managers. It seems as if the idea of National Service Farms ignores the experience in mechanized farming in Tanganyika up to now.

Villagization, the most important approach to rural change, must be subject to economic and psychological considerations. There are no local institutions able to carry out land consolidation and the planning of future land use. It is doubtful whether an outside office entrusted with the task will find co-operative peasants. The idea is suspect to them. They are tied to the land farmed by their ancestors. Peasants like to live on their land, close to their crop. They are afraid of theft on faraway fields. Ambitious individuals consider being incorporated in an obligatory rotation system a handicap to the development of their economic talents. Mistrust can perhaps be overcome by offering substantial economic advantages, such as good water, schools, a dispensary, tractors etc. This, however, will make villagization extremely expensive. These services may come later, as a result of economic benefits due to villagization. As a start they are financially intolerable. As far as already settled areas are concerned, villagization is financially feasible only where almost all of the changes, and the work they involve, are carried out by the peasants themselves. This, however, is unlikely.

In view of the difficulties which can be expected with villagization in settled areas, trial schemes are to start on newly cleared land. These schemes are actually "planned co-operative settlements". They differ from the above-mentioned "planned settlement" in so far as participation is to come from particularly active farmers, not from the urban unemployed. Those taking part would be expected to settle in a village. Thus villagization is part of the opening-up of new land. The success or failure of these schemes will depend on whether there will be a clear organizational set-up with "production under close supervision" and whether experience so far gained will be used.

e) US Maize, a Danger to Self-Help

Maize contributed by the US has played an important, partly helpful and partly dangerous role in settlement and self-help schemes. 90,000 tons of maize and 2,000 tons of powdered milk were supplied between 1961 and 1963. It began because of failing crops. In the beginning maize was supplied unconditionally to the needy. Later on, the distribution of maize was combined with emergency projects. In January 1962, 48,265 adults were registered for 207 emergency projects[1]. One adult received for one day's work 1.5 shs plus 1 lb of maize and 2 oz. of powdered milk. With the passage of time allocation of maize has become an important attraction for settlements and self-help schemes, so that it is not always easy to determine the motive for participation.

There is no doubt that US maize prevented starvation in certain areas. However, it seriously affected the willingness to work in others. In some places the peasants consider it more advantageous to participate in emergency schemes — where work is easy and supervision scarce — than to cultivate their own fields. Where maize has been distributed, cultivation has frequently been reduced[2]. It is noteworthy in this connection that much of the maize went into the Eastern Region, an area which is thinly populated and contains large tracts of comparatively fertile land. Local politicians were in a tight spot. People knew that maize was distributed and they exerted corresponding pressure on their political representatives.

f) A Change in Agricultural Administration and Extension

Self-help schemes and co-operatives are in the foreground of discussion. They are the forerunners of African socialism. They are at the heart of TANU and the government. Nevertheless, most agricultural development work is continuing according to the principles laid down by the British in the late fifties. On the one hand, this work has suffered due to the loss of experienced officers. On the other, extension is — due to political independence — meeting much more open-minded peasants. As to the effectiveness of extension, it is assumed in some locations that positive and negative changes keep the balance.

There is, however, a definite change from extension to administrative work. This is particularly true for the Agricultural Regional and District Office. Much time, money and enthusiasm is lost because officers are compelled to satisfy local desires for schemes which have little chance to be economically reasonable. The desk work of agricultural officers is accumulating:

— setting up of detailed development plans for districts and villages;

— planning and organization of settlements;

---

[1] According to Newsweek, December 24th, 1962, one million Tanganyikans received US food. The number of registered adults in May 1963 was 45,000.

[2] A report from Same District, Tanga Region says about certain areas: "People have stopped farming and live on fishing and US maize. There is no doubt that where US maize has been distributed, less land is being cultivated. In some locations this is creating the conditions for another famine.

— checking applications for loans. A rain of money was poured over the peasants in 1962. Loans should be made available for the good but poor farmers. Thousands of applications came to the District Office.

Also, there is a gradual change in extension policy. The principle of concentrating on interested individuals, active tribes and locations with good markets neither fits into the idea of a "levée en masse", nor into present interpretations of African socialism. Extension work is to approach the group, the community. Commercially ambitious farmers, it is said,

— become rich at the expense of others (which is untrue),
— keep their knowledge to themselves (which is largely true),
— are not interested in their neighbour's progress (with is largely true),
— shy away from communal enterprises (which is largely true),
— pay wage workers poorly (which is true).

Therefore they should not be specially promoted through extension work which is paid for by the general public. Individual achievements should not be used as examples of extension work, but successful communities.

There is some doubt whether the new approach will work as well as the old one. It is unlikely that small loans, which have been distributed so suddenly, will be repaid in time. The mistrust of commercial farmers who employ seasonal labour is short-sighted. They employ those who do not employ themselves. Approaching everybody means including those interested as well as the lazy ones. This amounts to a waste of time of expensive extension workers. It is inevitable that agricultural extension must concentrate on active individuals and active groups if it is to show a profit in Treasury terms.

On the other hand there are promising adaptations. Here and there extension workers use the Village Development Committee as their agency for disseminating innovations. This approach can be observed particularly where land is scarce, where higher yields are the only answer, and where the Agricultural Officer is the man to show how yields can be improved. There is the idea to make the Agricultural Field Assistant the most important figure in the Village Development Committee and to make him propagate there the policy of the Agricultural Department. The principle of concentration is still alive indirectly. Qualified personnel are placed for preference where the best development chances are.

All in all, however, the hardly begun period of systematic agricultural extension as the main job of Agricultural Officers is more and more given over to agricultural administration. This tendency is likely to continue. Agricultural extension is a western concept based on the assumption that isolated extension workers act on their own initiative and judgement, and that they are trained to adapt innovations to local circumstances. Successful extension requires patience. Sometimes a long period has to elapse between the introduction and acceptance of an innovation. Yet the leaders of a newly independent African country are unlikely to have the necessary patience. They need rapid, measurable and visible results. Many of them think in terms of ordinances and controls, they tend to overlook the difficulties which can be expected when applying controls in a country as vast as Tanganyika. Consequently, agricultural extension work as handed over by the British is a strange conception to many of the people in charge today.

g) Stagnation in Land Tenure

Another consequence of political change is the stagnation in land tenure. The British contributed to the revolution in peasant farming in Kenya by introducing the private ownership of land. They attempted to individualize land ownership in Tanganyika according to local wishes. As independence approached, this activity slowed down. The solution to this central problem, which is a politically explosive one, was to be left to the new government. The views of the new government are evident in this statement made by NYERERE:

> "... we must reject the individual ownership of land. To us in Africa land was always recognised as belonging to the community. Each individual within our society had a right to the use of land, because otherwise he could not earn his living and one cannot have the right to live without also having the right to some means of maintaining life. But the African's right to land was simply the right to use it, he had no other right to it, nor did it occur to him to try and claim ... The TANU Government must go back to the traditional African custom of land holding. That is to say a member of society will be entitled to a piece of land on condition that he uses it."

This formulation leaves scope for a lot of interpretations. The transfer of individual rights to individual farmers can easily be connected with conditions which ensure cultivation, and prevent the spread of tenancy. Yet this statement from NYERERE must be interpreted as implying that hopes are being placed on co-operative types of land use. Beginnings in this direction should not be made more difficult by granting more land rights to individuals. In consequence, a change in land ownership — that powerful lever in agricultural development — is delayed even where the peasants would welcome it.

h) The Estate Economy and African Socialism

It is understandable that estate owners and managers look with concern at the intentions of the new government which is striving to bring about social change and political consciousness. The estate economy needs social peace and secure economic and political conditions, particularly since investments in sisal, coffee, tea and ranching have production cycles of at least 10 years. The declaration that Tanganyika is a socialist country, or at least is to become one, is interpreted as a threat to private property. No clear definition of "African socialism" has yet been made. It is true that leading politicians declare that the new government will guarantee the security of foreign investment. Minister of Economics KAHAMA, has stated that there is no question of nationalization. Where there is nothing, nothing can be nationalized. Yet estate people hear of speeches made by lesser functionaries or labour union officials before African audiences, where rather different views are expressed.

The change in the tenure of estates from "freehold" to "leasehold" has caused concern. Rights on land which were established during the German administration have been changed to 99-year leases. This measure actually changed little, except to include the justified statement that rights to land presuppose that the land will be put to use. Yet there is concern about the order that agricultural district officers have to check land use on estates. Increasing wages and taxes add to uneasiness. An export duty on sisal and tea came into force in 1962. It amounts to 5 per cent of the export value of sisal, which is priced at over £ 60/ton. For tea it amounts to 3 per cent of the export value. Additional local taxes are levied, and a substantial turnover tax is charged for the establishment of Marketing Boards. New living quarters for workers have to be built in many places. Thanks to favourable prices for sisal and tea and to the technical progress in coffee these additional burdens can generally be carried without difficulty. The fact that the government itself is the second most important employer helps to control wages. New laws give the government extensive powers to arbitrate between employees and employers.

Nevertheless, the estate economy takes up the attitude of "wait and see". White settlers have less confidence in the future than commercial firms, and there are reasons for it. The politically conscious African has a special dislike of white settlers. Commercial companies employ personnel who come and go. Africans can advance into leading positions. The white settlers on the other hand lead a way of life which recalls colonial times. They want to establish permanent homes, while the Africans prefer to regard land use by foreigners as a temporary necessity.

It was expected that independence would bring changes in the estate economy. It must, however, be noted that efforts to mobilize the masses and to bring about social changes, add to the uneasiness of the owners and managers. It has had little influence on current operations, but has certainly affected investment and, consequently, expansion.

IV. The Principal Objective: Change in Social and Economic Attitudes.

This presentation of the institutions, achievements and difficulties should not be used to measure the success or failure of the "People's Plan" [1]. For the present the intention is much more to bring about a change in social and economic attitudes than to fulfil actual plan targets. Colonial rule is held responsible for the fact that there are so many "uneconomic men" in Tanganyika. Activities such as the self-help schemes, in which hundreds of people "tread on each other's feet", are uninteresting as far as material achievements are concerned. They are meant to show that social circumstances have changed, that responsibility is no longer in the hands of a small number of foreign officials, but is held by the Africans themselves. The Ministry of Co-operative and Community Development states:

> The objective of the people's plan for development is more than initiating thousands of village schemes; and more than the physical and tangible achievements in terms of village roads or buildings, impressive though these are now, and will be in future. It is not the physical, visible results which are the main achievement; it is the changes which take place in the people while they are undertaking the planning and execution of village projects.

The Minister stated before Parliament on June 25th, 1962:

> We are not merely concerned with the effort which the people are putting forth in the way of building village roads, digging wells, or setting up literacy groups, but with creating permanent organizations among people at the lowest level, through which they can think for themselves about how to overcome their problems, take action and, in the process, become more self-reliant and responsible citizens.

One might very well ask whether this objective is met by the approach chosen in the first year of independence; whether the activity of inexperienced young men will not deter economically active farmers, whose con-

[1] The same is true for production figures. Success or failure of efforts to increase output cannot be judged by the data from one or two years. 1961 brought a poor crop after 6 good years. Low output in 1961 was definitely not due to a reduction in British activitities because of the approach of independence. 1963 is likely to bring a bumper crop. It is possible that political efforts to increase acreage have had a share in this. The weather, however, was extremely favourable. Even fields planted late showed good growth. There are other factors which have to be taken into account. Poor crops in 1961 and 1962 led to high prices. There was a substantial price incentive for farmers. In addition, certain speculations favoured increased cultivation. It is very likely that two poor years are followed by a good one. Consequently farmers are said to consider it worthwhile to put more effort than usual into cultivation.

servative attitudes are proverbial; and whether the continuation of development policies worked out by the British would not achieve better results. Time will tell. Present observers may note, however, that unpaid labour is in fact being performed, that more land is being cultivated, and all of this in an atmosphere which is definitely not hostile to foreigners.

Even though one may assume that the British approach in the late fifties was sound and effective, the question remains unanswered whether it is politically sensible and feasible simply to continue a policy which was devised by former foreign overlords. Perhaps there must be a clear-cut break with the past in order to attain the change of mind which is required for indigenous economic development.

V. Thoughts on Future Development Policy

The attainment of independence is usually characterized by rapid changes. There are already signs indicating that future development efforts in Tanganyika will not be limited to an appeal for self-help on the one hand and to the three-year development budget on the other. It is quite conceivable that agricultural policy will change in three stages.

1st Phase: The Honeymoon of Independence.

The first years of independence brought great expectations, hasty changes in personnel, and an appeal to African patriotism. Government and TANU avoid unpopular measures. Activity was encouraged rather than ordered. The rural population should work out its own development targets. Those projects are undertaken which are spontaneously chosen by the people rather than those deserving priority.

But an agricultural policy motivated by such considerations leads to the difficulties mentioned previously. In addition, there is some erosion of state authority. Political independence is expected to do the impossible: to do away with the per capita tax, to nullify ordinances prohibiting settlement on forest reserves, to do away with all agricultural regulations, to create higher producers' prices, etc. Slowly people are beginning to realize that these expectations cannot be fulfilled. However, the peasants are even less willing to conform to regulations than they were under British administration. People settle on private or state forest reservations and clear the forest for cultivation. Cotton stalks are not pulled up in time and burned, as the regulations prescribe. Insects thrive and cause losses which are estimated at some thousand tons of cotton. It seems even less possible than before to enforce any kind of regulations, to make the peasants pay for irrigation facilities or for veterinary services. Local officials, Africans as well as British, are not sure that they will receive support from their superiors if they take decisive action.

The difficulties of early independence, the fact that there are frictions, tensions and complications need not be regarded exclusively as a disadvantage. Unrealistic small schemes which fail may be considered an unavoidable investment in experience. Many officials and functionaries who were unprepared for positions of economic and administrative responsibility have gained some understanding of the problems involved in economic development, and this was accomplished without serious economic losses. NYERERE maintains that the first task is "to get the country to face reality and truth". This occurred to some degree in the first year of independence.

2nd Phase: Introduction of Ordinances and Controls?

Fortunately the realization is growing that without governmental authority, a disciplined staff, the backing of subordinate personned, etc. no policy for economic development can be carried out. Consequently there are indications that another phase is being prepared. Perhaps the peasants are again to be subjected to ordinances and control, as they were in the early 1950's. The remark is heard that "the honeymoon of independence is over". Following are some indications of a "tightening of today's loosely held reins".

— An "Act to make provision for controlling and regulating the production, cultivation and marketing of agricultural products" was passed. This act gives the Minister of Agriculture absolute authority. He can prescribe what, when, where and how much to plant, to deliver, and to process, and what prices are to be paid. It is unlikely that much of it can be enforced in the near future. The legal basis for enforcement, however, has been laid.

— The Government is establishing Marketing Boards for products which until now have been trated freely. A central Marketing Organization is being prepared which is to dispose of such power as to make possible "the enforcement of the government's economic aims". The co-operatives will have to trade in agreement with this state marketing organization.

— Village Development Plans may be regarded as a first step towards obligatory cultivation and perhaps even towards obligatory delivery of production targets. For the time being the cultivation plans are being left to the villages — in so far as the discussion of such plans actually reaches village level. Village Development Councils are frequently setting high, unattainable targets. Non-fulfilment offers a suitable approach for more pressure. Officials need not say "you must plant so many acres", but can ask, "why haven't you planted what you promised to plant?" and then impose cultivation targets which have to be fulfilled, reasoning thereby that these targets will help to realize the local wishes as described in the Village Development Plan. Perhaps in the course of time — with the combined work of the Community Development Department which encourages and explains, and the administration, which enforces — unpopular measures can be enforced, such as the pulling up of old coffee trees, burning of the cotton stalks at the proper time, early planting of cotton, reduction in the number of stock, terracing, etc. There is among TANU secretaries and officials a significant willingness to press the peasants. Reference is often made to the period of German administration, of which it is said that rapid progress was made, due to the unchallenged authority of the administration. The British method of "persistent persuasion", concentrating on interested individuals, and avoiding direct compulsion is strange to many leaders at the local level. In some places local officials are just waiting for backing from Dar-es-Salaam in order to enforce, if necessary with compulsion, what they consider essential for development. In other places, however, the opposite is true. Measures which might be unpopular are studiously avoided. The attitudes vary from place to place, as can be expected in a country as large and as manifold as Tanganyika.

— The emphasis on villagization and co-operative settlements similar to those of the Moshav System of Israel can also be interpreted as a step towards more controlled and guided production. Also there is a growing realization that the responsibility for irrigation facilities cannot remain in the hands of unskilled peasants. In future, it is said, the managers of settlements and irrigation schemes should no longer be obstructed by local people.

— The desire for rapid economic development leads to a search for finance. Attention is focusing on those "100 million pounds of capital in livestock" which are owned by peasants and herdsmen and which bring in very little return. Re-introduction of a tax per head of livestock is discussed. This would bring about: (1) additional tax revenues, (2) the rationalization of animal husbandry, i.e. increased production of meat, and (3) less attraction to invest earnings from cash crops (coffee, cotton, pyrethrum) in livestock. With approaching independence, the tax on livestock fell in to abeyance because it was an extremely unpopular measure. Recently the question has come up again.

3rd Phase: Return to a Selective Approach?

Although the realization of the necessity for discipline, control and "pressure situations" must be welcomed, the question is whether the government of Tanganyika is in a position to enforce extensive and far-reaching controls and regulations. Ordinances enforced by compulsion were an important component of colonial development policy. The result, however, was more often than not a failure. The presence of a one-party government, the fact that compulsion is being applied by Africans instead of foreigners, and that not only compulsion but also an appeal to African patriotism is being used, may induce peasants to be more willing to conform to ordinances. On the other hand, one must not overlook the fact that the control of agricultural production is more difficult in Tanganyika than in almost any other country, in view of the great distances, poor communications, scattered settlements, absence of cadastral surveys, shortage of trained personel, etc.

It is quite conceivable that trial and error, accelerated by a shortage of money, will bring about a selection of those approaches which have already proved successful in the past, viz. concentration of extension on interested groups, individuals, villages and tribes, supplemented by co-operative settlements with "production under close supervision" in some carefully chosen cases. In short, it is not impossible in the long run that some essential components of British agricultural policy in the late fifties will be reintroduced step by step, simply because they are appropriate to the situation.

VI. Conclusions

Considerations on future development policy are largely a matter of speculation. The fact that a basis has been laid for ordinances and controls does not necessarily mean that they will be introduced. One thing however is quite clear: agricultural development in Tanganyika in the near future will not be a mere continuation of the British approach. Expedient as it may have been from a technical and economic point of view, its success was dependent on the presence of numerous European officers in responsible positions. This condition proved to be no longer valid in the psychological situation in the early years of independence. Tanganyika wants to go its own way.

One may criticize this or that approach in the agricultural development policy of the new government. The basic idea, however, is right: increasing production through self-help, through better exploitation of resources held by the peasants, without major capital investment in farming. Under conditions as they are in Tanganyika, agriculture can largely be developed by cheap means i.e. by activating the peasants. Available capital should go for preference into urbanization and industrialization, and should thus create the internal buying power essential for agricultural progress.

It is probably appropriate to Tanganyika's present situation that experiments in agricultural policy are being made, provided these experiments are not too expensive. Learning by trial and error is an essential feature of developing indigenous structures. It is difficult to say how long this period of trial and error will last. The apparent pragmatism and desire to achieve results may lead to the rapid adoption of methods which work. In any case, those interested in Tanganyika's economic development must take in to account the fact that there is bound to be a time of ferment. This is not a pessimistic statement. There is some, but not extreme danger in a period of fermentation, considering that most of the people live on the land, that they are quite content with their present situation, and that subsistence is generally assured.

In the long run, present efforts to "change minds" may prove to have been the right approach and may lead to economic progress, provided that:

- — African socialism remains so pragmatic that the government is free enough to chose in each case the most efficient approach;
- — there will be a build-up in discipline, puritanism and devotion to the common cause.

It can hardly be doubted that the last-mentioned point is decisive.

Chapter F

Agricultural Development Aid: Some General Observations Based on Conditions in Tanganyika

I. Some Fundamental Points

1. Aims and Dangers of Development Aid in Agriculture

To be able to answer the question whether and in what form agricultural development aid under conditions as they exist in Tanganyika would serve a useful purpose, requires a clear definition of the aims in view. What should be the purpose of such aid — humanitarian, social, commercial, or political? The economist can deal with the matter concretely as soon as development aid is given for the purpose of "economic development".

This concept is a collective term for a number of interrelated processes. Economic development means the use of new energies and new methods of production [1]. Structural changes take place. Agriculture and the village lose in importance in relation to industry, trade and the town. The production of capital goods grows more rapidly than that of consumer goods. The division of labour, technological progress and capital formation lead to greater productivity of labour and, at the same time, to a higher standard of living and yet more capital formation. It is customary to measure this process by the rate of the per capita social product. The complex process of economic development cannot, of course, be gauged accurately by this one criterion alone. This is not the place to discuss the many difficulties arising from the employment of such a yardstick. What is needed is a comprehensive scale: for lack of a better we are compelled to use this one.

We can, therefore, define the aim of development aid as follows: high and continuous rates of increase in the per capita social product. In view of the special conditions existing in Tanganyika we would add "with the Africans taking an increasing share of responsibility". Continuous and high rates of growth of the social product necessitate:

— rapid adoption of technical, managerial and social innovations, and
— high rates of capital formation.

---

[1] For a definition of the concepts "economic development" and "industrialization", see Klatt, S.: Zur Theorie der Industrialisierung. Cologne/Opladen 1959, p. 19 ff.

Hence, if economic development is the aim, it must be the function also of agricultural aid — as part of the whole — to serve both these purposes.

At the same time other purposes of agricultural aid frequently mentioned can be pushed aside as not corresponding to the primary aim, or as having to be included in a more comprehensive structure.

Fight against Hunger. There are no comprehensive estimates of the average intake of calories in Tanganyika. This is generally taken to be adequate. Almost everywhere in Tanganyika families are able to produce the food they need without excessive efforts or additional means of production. Chronic cases of hunger are the result of either social hardship — disease, widows with many children, etc. — or apathy. The demand for a "war on hunger" has, therefore, little relevance in Tanganyika in respect of the supply of carbohydrates or the opportunities to produce them. Under these circumstances supplying food cannot be considered as development aid and only in exceptional cases as "aid on humanitarian grounds". Self-help can be expected only when the peasants know that they will have to go without if they do not grow sufficient quantities of food.

Occasional bad harvests, however, and inadequate nutrition constitute real problems.

The Problem of Famines. The problem of occasional famines cannot be solved by gifts of food or by general measures to promote agriculture. What is required are structural adjustments:

- *Migration* to areas with adequate and certain rainfall, i.e. the raising of tribal boundaries.
- The development of *animal husbandry as a "cash crop"*. The chronic distress areas of Tanganyika are in the main those where the people have plenty of cattle and the land is suitable for "ranching". A rational use of herds and grassland could produce an income large enough to allow the purchase of additional food in case of bad harvests.
- *Enforced cultivation* of manioc and sorghum. In recent years maize has increasingly taken the place of manioc, sorghum and millet. Doctors recommend the eating of maize, as it contains valuable proteins and fats. However, maize is subject to greater yield fluctuations. Manioc should be planted as a complementary crop. The farmers, with their own brand of lethargy and optimism, are given to speculating, hoping that "their maize" will "this year" get enough rain. In view of this, an administratively imposed ordinance to grow manioc on a minimum acreage per community would be a cheap way to mitigate poor harvests.
- The installation of *irrigation* works is undoubtedly a splendid way to prevent poor harvests. It is very costly, however, and necessitates shifts of population, such as from the arid Central Province to the Rufiji Basin.

These structural adjustments would not, of course, solve the problem of an occasional bad harvest altogether. It does sometimes happen that the harvest dries up, even in areas with usually regular rainfall. But if these adjustments were carried out systematically, one assumes that the number of famines would be greatly reduced.

Qualitative Improvements in Nutrition. The problem of quality in nutrition is less of a technical or economic nature than one of social attitudes and education. The peasant families practically everywhere in Tanganyika have opportunities to obtain an adequate supply of animal proteins and vitamins (milk, goats, poultry, fish ponds, vegetable gardens, fruit, or division of labour and additional purchases). The fact that the existing opportunities are not fully made use of is due less to any objective difficulties than to the subjective behaviour of the people and their fixed attitudes. They are not in the habit of looking for improvements or innovations. It would not be sensible to change the situation by outside supplies of protein and minerals. Nor could lasting changes be expected from extension activities about nutrition. Greater success could probably be achieved by drawing farmers into the exchange economy, which would gradually lead to a change of mind.

Erosion Control. "Erosion control" is yet another slogan which could lead to agricultural development aid coming to nothing. It is true that there are areas where one can find vast damage due to erosion. Tanganyika is still regarded as relatively well conserved. But here, too, are large areas where the vegetation has been almost completely destroyed. We do not deny that long-term erosion control is inevitable. Lell certain is the stage of economic development at which major efforts in erosion control should be made. In Tanganyika, the answer to this must be negative for the time being. Erosion control is an expensive business. Even with considerable means one could do no more than a job of patching up. It may be assumed that erosion control on any comprehensive scale is possible only with the funds of a developed industrialized society. There is, of course, the possibility of using unpaid labour in order to control erosion. One can expect the population, however, to oppose such measure with all possible strength. Again, the more successful way of controlling erosion is probably the commercialization of the farms. Sooner or later this raises the question of manuring and crop rotation and, therefore, of soil conservation, plus the fact that it gives the farmer the means — even though this may take some time — and the insight to change the situation.

Monoculture. The objections against monoculture have to be treated with the same caution as the demands for erosion control. In many instances monoculture is necessary, if there is to be any successful farming at all. Monoculture is not always harmful to the soil. The most common monoculture, the farming of paddy rice, may on the contrary be regarded as a classic example of a soil-conserving form of cultivation. Similarly, from the point of view of soil fertility, one cannot object to a monoculture of tea, coconut, palms, sisal, etc.

At the beginning of any economic development monoculture is an economically unavoidable form of cultivation. Variety in farming is expensive and requires additional investment, costly supervision and a market for the various products. In view of the marginal production situation, this is truer of Tanganyika than of many other countries.

Raising the Rural Standard of Living. It is undoubtedly true that it is a final objective of economic development to raise the rural population's earnings and standard of living. But a continuously higher standard of living is possible only if capital formation is given priority over consumption. Since high rates of capital formation on a national basis are essential, both to safeguard future economic growth and to achieve economic independence, it seems proper to base agricultural development aid and policy on the possible effects on capital formation rather than on raising the living standard. This is particularly true in a country like Tanganyika where the greater part of the population does not regard the present living conditions as unbearable. There is very little obvious poverty of the kind one may encounter in many places in India for instance. In these circumstances attention should be paid less to raising the standard of living than — as far as is politically practicable and socially justifiable — to securing the future by way of capital formation in every sense, i.e. both in factories and the infrastructure and in "human capital", i.e. in knowledge and economic attitudes.

The examples mentioned make it evident that it is not advisable to offer development aid with specific technical or social aims in view. There may well be instances when it would be opportune to promote measures of erosion control or to raise rural consumption. Yet the yardstick is not technical achievement or the raising of consumption, but the question: to what extent did this measure contribute towards stimulating cumulative growth?

When distributing development aid, it is also necessary to be aware that it might well result in delaying economic development. Decisive for the success of development efforts is not the supply of money and personnel, but the spreading of an active and positive attitude towards the economy. Without such an attitude, which is something development aid cannot create, but which the recipient country itself must bring about, even the greatest amounts of aid would be wasted. This applies to agriculture even more than to any other field. One can assume that the essential change in the way people think, particularly those in responsible positions, will occur only when it is a matter of dire necessity. This dire necessity is often the result of the tension between the resolve to get ahead and the lack of means to do so. Among those in authority in Africa today there is no lack of will: what is lacking is experience. In these circumstances the lack of money can render an important service. It demands a quick sorting out of reasonable and impractical ideas. Reversely, development aid can lead to people adopting the attitude that "anything will do", in short, the country is harmed by necessary adjustments being postponed. Hence the purpose of development must be:

— that it will stimulate a cumulative development, avoiding the danger of "stagnation on a higher level";
— that it does not sap but, on the contrary, encourages the desire for self-help.

2. Agricultural versus Non-agricultural Development Aid

The demand that agricultural development aid must be subordinated to the higher aims of "continuous economic growth" raises the question whether, and if so, to what extent, agricultural development aid deserves to be given priority over aid in other fields. It is argued that agriculture should be given preference because it is the most extensive sector of the economy. This argument is by no means convincing. It could be said with equal justification that non-agricultural activities deserve to be given priority, exactly because there are not enough of them.

This question — what portion of development aid should be devoted to agriculture — is not so easily answered. In the case of Tanganyika the main arguments for and against agricultural development aid are as follows:

(1) In favour of agricultural development aid:

A Favourable Ratio between Expenditure and Returns. In peasant agriculture the opportunity exists to increase production fairly quickly without heavy investments, by a better combination of existing resources and the introduction of technical innovations. It is not possible to exploit this opportunity without development aid, particularly in applied research, training and extension work.

Domestric Purchasing Power. The main reason why agricultural development aid projects are given preference is the absence of profitable projects outside agriculture. The purchasing power of the population is low. It is not profitable to establish consumer industries if there is not a big enough market. First agriculture must be developed and this will raise the purchasing power of the rural population — so runs the argument — before there will be a profitable market for additional industrial projects.

Own Efforts. A particular advantage of agricultural development aid lies in the fact that it can be used as a stimulus to self-help. A kind of local leadership develops. The resulting products are regarded by the farmers as having been achieved by their own efforts. This offers good conditions for the development of a responsible relationship between man and material goods.

Variety of Aid. Economic development in Tanganyika is unthinkable without experienced Europeans and Asians. As long as aid is coming from only one source, the fear is that it will result in foreign domination or in a continuation of a neo-colonial dependence. It helps towards a harmonious co-operation between the races if development aid, in agriculture, too, therefore is derived from as many different countries and cultures as possible.

(2) Against agricultural development aid:

A Poor Market. Although Tanganyika's domestic markets offer some scope for more agricultural products to be sold at favourable prices, the total purchasing power of the consumers is not big enough to absorb a rapidly growing supply at prices conducive to further production increases. It seems unreasonable to further production with the result that prices fall sharply, so that the stimulus for more production increases is blunted.

Predominantly agricultural development aid is identical, therefore, with increasing agricultural exports. But this encounters a number of difficulties. The export of coffee is restricted by the quota laid down in the coffee agreement. Already there is the danger that the supply will exceed this quota. Cotton farmers are already subsidized. In the case of cattle, the occasional occurrence of rinderpest obstructs the profitable export of refrigerated meat. The prices of many other products are so low, or the costs of transport so high, that it would not pay to increase cultivation. Not all agricultural products have to contend with unfavourable export markets. The chances for sisal, tea, sesame etc. are quite good at the moment. But altogether it does not seem advisable to depend even more than hitherto on the uncertain and badly paid export of agricultural produce by giving priority to the promotion of agriculture.

Rural Unemployment. Predominantly agricultural development offers little prospect to the problem of employment. In this connection two aspects are of importance. One is that Tanganyika has so much labour available that agriculture and industry do not have to compete with each other; the second is the fact that predominantly agricultural development is accompanied by a high and continuous rise in the population. There is the danger that the high rates of increase in the rural population may cancel out any successes in the development of agriculture, if these are not accompanied by greater urbanization. To the extent that labour can be gainfully employed in the towns, the adjustment of the population increase to urban conditions must be regarded as a positive factor beyond the productive contribution in terms of labour.

Predominance of Traditional Thinking in Rural Communities. The traditional rural community incorporates a multitude of mechanisms designed to restore an equilibrium rather than to encourage cumulative development. The odd innovation may well be taken up; but the generating factor is small. Surpluses are used up in consumer investments: cattle purchases, house building, trips in taxis, etc. If agricultural development does not go hand in hand with urbanization, the danger exists that any initial successes will come to a dead end.

Agricultural development brings a new orientation of values, a rational treatment of soil, cattle and plants. This comes about more easily the more the village is influenced by the town and the economic attitudes prevalent there. In consequence, as soon as local markets are established with some purchasing power one can expect an almost autonomous development of agriculture, or one that can be brought about at little expense.

Few Secondary Effects. In addition, agricultural development has relatively little effect in other fields. In the wake of industrial investments other effects follow more or less inevitably. Industrial sites have of necessity a stimulating influence on a multitude of other activities. In thies respect agriculture occupies last place. For instance, the production of tea can increase considerably without this affecting any other activities [1].

Experiments in Agricultural Policy. A correlative to political independence is the search for social models. In Tanganyika, agriculture is the main sphere in which new ideas are tried out. Efforts are certainly welcome which develop an African identy and work out economic concepts with links in the tribal system of the past — they are given the general term of "African Socialism". However, in these circumstances one must expect agricultural aid to be partly used up in experiments. Quite apart from the fact that practical concepts have a habit of being put into effect more rapidly when there is little money available.

Shortage of Personnel. Agricultural development is more dependent on administration than industry. Success or failure depend in the first place on the quality, leadership and supervision of the staff of officials. For the organization of a large number of medium and small projects no authority other than the responsible ministry is normally available.

In view of the fact that the civil service has undergone rapid changes, for some time it must be assumed to be less efficient. For agricultural development aid to bear fruit it is, therefore, necessary for projects to be provided with personnel from the country giving the aid. This, in turn, involves disproportionately high costs for personnel. Industrial projects are less

---

[1] On this question, see particularly HIRSCHMAN, A. O.: The Strategy of Economic Development. Yale University Press, 1958.

dependent on administration and the government. Since as a rule these are rather large — in comparison with agricultural projects — it is probable that the ratio between output and expenditure for supervision is more favourable.

Dividing Tasks between Aid-Giving and Recipient Countries. Because of the great number of small agricultural projects which are difficult to control, it seems sensible to divide the tasks between the countries giving development aid and the recipients. Agricultural development entails direct contact with the population, a knowledge of the national language and the local customs, and securing the agreement of various administrations and the political party. It is the ideal sphere in which African leadership can grow in experience without too heavy economic losses. It is, therefore, advisable to finance agricultural development mainly from domestic sources, and to employ African personnel to see it through, while the donor countries should assume responsibility chiefly for larger projects concerning industry and the infrastructure.

Dividing the Tasks between the Various Aiding Countries. Furthermore, it appears reasonable to divide the tasks between the donor countries themselves in accordance with their respective specialized knowledge. Hence, the idea suggests itself that the German contribution should go towards industry rather than agriculture. Other countries, such as the USA, Japan, Israel, France, Great Britain, etc., have a great number of agricultural personnel familiar with tropical agriculture, the behaviour of villagers in the developing countries, and the handling of small agricultural projects. In Germany, there are few such people.

There are also practical considerations in favour of the tasks being divided within the complex of agricultural development aid. The USA, Israel, Italy, etc. have great experience and personnel trained in such subjects as irrigation, rice cultivation, and citrus fruit: in much the same way Germany could concentrate its contribution on subjects such as co-operatives, milk production, pig-keeping, mineral fertilizers, forestry, etc.

(3) Conclusions on the distribution of agricultural development aid.

The attempt to arrive at a practical policy for agricultural development aid in countries like Tanganyika from these basic considerations leads to the following conclusions:

Priority of Non-Agricultural Projects. Non-agricultural projects, especially in industry, should be given preference if they are paying propositions, or if they are economically of strategic importance as links in a production chain, or as development centres [1]. They create purchasing power

[1] HIRSCHMAN warns against priority being given to infrastructural investments, such as road construction. Most visitors to Tanganyika regard this as one of the most urgent problems. But it is unlikely that the building of roads would encourage agriculture to such a degree that it would constitute a more profitable undertaking than

for agricultural products. They help to concentrate this purchasing power in larger urban centres which one can presume to exercise an influence on the economic thinking of the surrounding rural population. Industries for the processing of agricultural raw materials help to improve market conditions, and serve at the same time as nuclei for the implementation of modern farming methods. For this reason one can regard measures helping in the development of industry and towns as the most important form of agricultural development aid.

Limits of Non-Agricultural Development Aid. It is unlikely that Tanganyika offers much scope for enough projects outside agriculture. The fact that the non-agricultural sector of the economy is not capable of absorbing a sufficient volume of aid leaves no other choice but to include agriculture in development aid.

Exploiting Marketing Chances. Agricultural development aid should be concentrated on exploiting specific marketing chances. It is not advisable to encourage production for the domestic market beyond the level of its purchasing capacity. The only export products justifying money being spent on their promotion are those for which a profitable market can be found without this clearly conflicting with the interests of other developing countries which are also dependent on development aid.

Limiting Aid to Schemes with a Favourable Input-Output Ratio. It is appropriate under the conditions existing in Tanganyika to limit agricultural development aid to schemes where the input-output ratio can be expected to be favourable. It is useless to establish expensive production: domestic purchasing capacity is not great enough. To subsidize agricultural exports is not a wise policy for a country without balance of payment difficulties. This does not mean that development aid should be devoted only to schemes, the production of which is competitive on the world market. *The point is that — if there is a need for subsidies — non-agricultural activities justify such protective measures more.*

Preference for Aid Stimulating the Farmers' Own Initiative. The best way to bring about increases in agricultural production at little cost is the rational use of the existing resources of a peasant economy by encouraging the peasants' own initiative. The agricultural policy of the British ad-

---

others. The idea that the people are only waiting to take advantage of production chances which new roads would make possible, presupposes an attitude of mind such as might be found in industrial nations, but hardly in Tanganyika. Road construction is not enough to unlock the potentials of an agriculture; one also has to activate the people through agricultural extension, additional investments, etc. The economic usefulness of new roads is frequently adversely affected by the shortage of funds for these secondary services. In the conditions prevailing in Tanganyika it would seem more important to concentrate on the fuller development of areas which, from a transport point of view, are already provided for, than on investing in further road construction.

ministration in the 1950's has shown that a profitable start in this direction can be made by the "persistent persuasion" of individual farmers, by specific projects of "production under close supervision", and by developing co-operatives. This means that there is a need for development aid in research, training, extension work, co-operatives, pilot farms, settlements, etc.

Agricultural Development Aid is Primarily Technical Aid. Hence agriculture needs primarily, though not exclusively, technical aid. This is again complementary to non-agricultural development aid which is chiefly a matter of capital aid. Again, it is decisive for the success of technical aid that help should be given in the form of personnel.

Altogether, keeping these principles in mind, agricultural development aid can be regarded as supporting a development policy with the long-term aim of industrialization and urbanization, but with the short-term objective of achieving the agricultural basis cheaply, as a prerequisite for progress in non-agricultural sectors.

3. Criteria for Agricultural Aid

The decision about an appropriate proportion between agricultural and non-agricultural aid leads to another question. What criteria should be used in order to judge whether certain agricultural schemes are worthwhile for development aid or not. The following points can be regarded as of cardinal importance:

a) Fiscal Profitableness

Of primary importance is the relation between input and output, whereby we mean by input the money and personnel invested in a particular project by an aid-giving agency and the government of the recipient country. The yardstick for output is not the gross return, or the farmers' incomes, but the additional revenues to the Treasury.

Conditions in Tanganyika are such that the farmers must be expected to use up increases in earnings or — perhaps even more important — to invest their savings in unproductive items, such as the "cattle cycle" already described. Without new impulses entering into rural life, the peasants' own initiative tends to be dissipated. Such stimulation costs money. With small returns to the Treasury the means of financing further projects are not forthcoming. If initial aid is spent without any cumulative effects it is nothing more than alms-giving. It is true that it is almost impossible to determine the degree of fiscal profitableness of research, for instance, or training and extension services. Nevertheless, there is no other way to decide the relative value of different measures, both within agriculture and in comparison with non-agricultural projects, but to estimate input and output in Treasury terms.

Making fiscal profitableness an essential criterion in the selection of development projects in agriculture by no means implies that only "profitable projects" deserve to be supported. The probability in Tanganyika is that there are not sufficient opportunities to engage in schemes whereby the Treasury can expect to gain an additional income by way of direct taxation, amounting to a rate of interest of $8^3/_4$ per cent on the capital invested — the interest rate used as a yardstick in the report of the World Bank. "Unprofitable" investments are hardly avoidable, if there is to be any economic development at all. This, however, does not invalidate fiscal profitableness as a criterion in the selection of projects. The point is that projects should be given preference which promise the best results in comparison with others.

b) Continuity of Stimuli

One can distinguish between development measures according to how lasting an influence they have. Following the point on fiscal profitableness within a reasonable time, there is the question of long-term effects. Training programmes can exercise a decisive influence on the future attitudes of leading personnel and must be judged by the extent to which they may be capable of effecting increased production in the more distant future. Continuity of effect is a matter deserving attention in the agricultural and technical sphere also. Schemes of irrigation, perennial crops, introducing dairy farming, etc. create lasting improvements, with others in their wake. On the other hand, the effects of annual crops are of shorter duration. There, some innovations are easily taken up, but as easily discarded in the face of small obstacles.

So-called "development nuclei" can contribute greatly to the continuity of agricultural development. Such a nucleus may be constituted from a group of progressive farmers, the whole population of a village alert to new ways, or maybe large farms with associated settlements under "close supervision". When such development nuclei serve to introduce new methods which bear on future developments, it may be sensible to help them financially, even if their direct fiscal profitableness is somewhat low. In such cases the profitableness would not lie in the project itself, but rather in its long-term effects as a major link in a sequence of interrelated activities.

c) African Responsibility

It must be assumed that under the present conditions in Tanganyika production by Europeans or Asians will not be welcomed, but rather accepted as unavoidable for the time being. This attitude endangers co-operation between the races and, therefore, hinders continuous economic development. To lessen this tension, which would increase if future activities in the economy were determined mainly by minorities, it is necessary to help the Africans to take over responsibilities. Any development aid which does not help the country to become economically independent — as to personnel also — does not conform to the political realities. Hence projects should also be judged on the grounds of whether they create capital in the hands of Africans, and to what extent they enable Africans to take over the responsibility for economic activities. To this end agriculture offers two kinds of opportunities: in the first place, the improvement of existing farms by way of agricultural extension services is in fact a long-term training programme for the farmers as well as officers; in the second, in the course of establishing modern farming nuclei, such as irrigation schemes, plantations, etc. special training programmes can be incorporated to prepare for gradual Africanization.

d) Social Pragmatism

Where the three criteria listed above apply, projects must further be judged pragmatically. It would be wrong and would lead to wrong decisions, if it was thought that the agriculture of developing countries must, or should, go through the same stages as in Europe or Japan — that, for instance, the change-over from hoe to ox-drawn plough, rather than to a tractor is more appropriate, or that the first step must be an improvement in the tools, and that complicated machinery such as combines should be left to more advanced stages of development. In many instances there are compelling arguments against the use of tractors. But whether, and to what extent, these apply should not be decided on the basis of vague theories, but according to the relevant technical, economic and social conditions. There is no comprehensive theory of economic development capable of answering such specific questions.

Nor would we recommend that development aid be offered for the sake of certain ideological objectives, such as "support for the independent middle class" or "promoting collective co-operation". Notwithstanding the pros and cons of such approaches, the attempt to exercise an influence in this or that direction would be registered as by seismograph, and would create suspicion and probably have the opposite effect. The country extending aid should choose only the approach which promises technical and economic efficiency. This would probably lead to an agricultural development policy comprising both support for independent farmers and for state-owned farms and co-operatives.

e) Political Considerations

The criterion of economic efficiency for development aid should be applied with some degree of generosity and far-sightedness. It is useless to engage in projects which do not meet the wishes of the country or its government, even though they may be economically promising. *Projects must be of a kind in which the people of the country concerned can put their hope and confidence,* and which are part of a development programme or plan. Indeed, the occasion could arise where it might be advantageous to include among the whole complex of aid measures, side by side with projects undertaken for predominantly economic reasons, some others, the primary importance of which is not their economic usefulness but the speeding up of the process of social change by way of symbolic "national tasks". It might even be advisable — within limits — to give financial support to some project which, in economic terms, is likely to be a failure. A country doling out development aid and stipulating in a schoolmasterly fashion how it is to be used, is bound sooner or later to evoke a feeling of dislike rather than gratitude. A measure of failure is necessary to the realization of practical concepts. Independent thinking is not acquired by just following the advice of others. As long as these failures are confined to relatively small experimental projects, the financial loss involved can be regarded as useful investments in the nature of a "experience capital".

f) Institutional Suitability

We must further distinguish between projects and measures important and necessary for economic development and those suitable for development aid.

In addition to the above criteria, a donor country is likely to assess this suitability according to whether

— a project reflects the aid given in an unequivocal and lasting manner. This is in the interest of the recipient country too. The amounts of aid given depend last not least on the politicians of that country being able to demonstrate to their voters the obvious achievements of development aid.
— the project can be carried out with little organizational difficulty. Projects needing the collaboration of a large number of local authorities and offices are suitable only if these function properly.
— much or little personnel is needed from the donor country. Generally speaking, donor countries are not inclined to employ a large staff in development aid. The fewer personnel that have to be dispatched the more the project is suitable.
— the personnel from the donor country are placed in positions which expose them to tensions in their relations with their African colleagues. Scientists or teachers are in less danger of coming into conflict than advisers or even managers.
— the project concerned would favourably affect the demand for goods manufactured in the donor country.

By reason of these considerations non-agricultural projects are as a rule more suitable than agricultural ones. They are relatively large, fairly independent of local authorities, do not interfere with people who wish to go on living in the old traditional way, and stimulate the demand for machines and other means of production which can be supplied by the donor country. In the field of agriculture it is research, training and extension which can be regarded as particularly suitable.

II. Particular Spheres of Agricultural Development Aid

Having stated the aim "to contribute to continuous economic growth" and the above-mentioned criteria for selecting projects, we will now discuss the chances for, and obstacles to, specific measures of agricultural development aid. It is not our intention to make specific proposals; the aim is to outline the points of view which should be borne in mind when judging particular proposals [1].

1. Agricultural Research

The rapid spread of technical progress in peasant farming is of key importance in developing agriculture "cheaply". To continue with the necessary tasks of applied agricultural research is unthinkable without development aid. The number of African scientists is as yet very small. In view of the long time necessary for scientific training, it will be decades before there will be a sufficient number of them able to carry on the work of existing research institutions. It is in line with the country's situation that educated Africans are directed to responsible positions in the administration and in industry rather than to research work. This will have to rely for quite a long time on foreign personnel.

There is less need for more research institutions than for the existing ones to be maintained and fully staffed in quantity and, especially, in quality. The following arguments speak against increasing the number of research stations:

— Independence means a change of employer. In place of the British Colonial Administration, which meant security and a chance of promotion, there is the government of an independent country. This encourages the tendency for personnel to leave. Development aid is necessary to guarantee that existing research institutions are maintained and, at the same time, that many years of experience are not lost.
— Agricultural research is bound to certain localities, in consequence of which research stations are rather spread out. This, in turn, implies loneliness and a lack of opportunities to exchange ideas. Scientific work makes it necessary for research people to be in touch with their colleagues

[1] The arguments are limited to the sphere with which the author is familiar, i.e. agricultural production.

in allied disciplines. It would seem advisable, for the purpose of engaging in additional research work, to extend existing stations, in order to turn them into more effective and attractive centres.

— There is hardly a country giving aid which possesses personnel familiar with the agricultural conditions in Tanganyika. Each new beginning requires the gathering of experience and some setbacks. The acclimatization of new personnel is very much easier in existing institutions than in new ones.

— If the existing institutions are fully manned and, perhaps, extended and reconstructed, they can be regarded as adequate for the needs of Tanganyika.

Another aspect of the economics of scientific aid is the question of what research projects should be selected. The generalization that research deserves support often disguises the fact that there is a great deal of waste. Within the framework of the project being divided among different countries, it would seem sensible for the industrial countries to take over the costs of basic research. The government of Tanganyika cannot be expected to be greatly interested in this. Because of the urgency of economic development, only such research projects can be regarded as necessary, the results of which can be applied in practice, either by advising the farmers and estates or by production "under close supervision". Many subjects of agricultural research rightly studied in industrial countries, are, for the time being, just not relevant in a country at Tanganyika's stage of development. It must also be considered that research requires an apparatus for disseminating its practical findings. Hence the question as to whether, and to what extent, agricultural research should be given development aid depends also on the efficiency of the agricultural extension service.

First in importance among the various branches of research is plant breeding. It is common experience that, at the beginning of economic development, there are marked successes in plant breeding. Tanganyika is no exception. Much has been achieved in the breeding of cotton, wheat, manioc, and sesame. In the field of animal husbandry, the health of the stock must be given prior attention. In the case of stock breeding and feeding, the results of research cannot yet be applied, except in some isolated instances. Another subject for research, hitherto hardly touched upon, is agricultural economics. What is needed are data on the current methods of land use and marketing, as well as on the input-output ratio of certain approaches in development.

As a rule it will hardly be possible for a donor country to decide which research projects should be supported. This would require a thorough knowledge of development aims, harmonizing with research projects in other fields and of other aid-countries. In view of this, it would seem desirable to co-ordinate agricultural research and research aid within the East African Common Services Organisation. This institution should submit proposals for research which might be undertaken by aid-giving countries.

On the question of personnel, it would be better, but not necessary, if the scientists sent out are experienced in tropical agriculture. It is more important for them to have a solid basic knowledge as well as a fluent command of English, and to be willing to learn from the experience of others. If the work is to be of any practical use, one would have to prepare for a stay of 5 years' duration. Individual studies, such as doctor theses, which are a matter of education rather than scientific experience, would require a stay of at least two years' duration. The success of newly arrived scientists depends in each case more on calibre and adaptability to local circumstances than on experiences in other parts of the tropics.

The necessity of such a stay being of some duration raises the question as to whether the required personnel would be available. Qualified scientists are generally not prepared to enter the service of the Tanganyikan government. There is little scope for advancement, such as existed previously in the "Colonial Service". A solution might be the example of the "Empire Cotton Growing Corporation". This research organization employs and pays scientists, and sends them on request to developing countries. Its services are paid for either by the respective country itself, or by one of the aiding countries. This large organization offers the scientists security in case of political entanglements, as well as a chance to advance professionally. Another possibility might be for research institutions in donor countries to take over on a permanent basis the responsibility for similar research stations in Tanganyika.

2. Agricultural Education

Like agricultural research, training, too, is a field particularly suited to development aid. Tanganyika has few African teachers in agriculture. In the course of Africanizing the agricultural administration some of the best of them have been transferred to administrative positions. Hence agricultural training will have to rely on Europeans for some time to come. And they are decidedly welcome as teachers.

Agricultural training can be promoted by way either of scholarships for study in the donor country, or of training centres in the recipient country. The first of these alternatives is by far the most popular, but the second deserves to be given preference [1].

[1] This opinion is based on economic considerations. The granting of scholarships is in practice a political act. Countries granting them usually hope to influence the scholarship students and through them the social development of the developing country. It is, therefore, in the interest of the developing country itself to prevent such attempts by building up its own educational system.

— To have to learn yet another language — frequently the fourth, besides the tribal language, Swahili and English — may be of value to a scientist. But in the case of a future agricultural officer it is an unnecessary and costly burden.
— The agricultural conditions in the donor country differ from those in the developing country. The student learns the technique applying to crops which are not grown in his own country (e.g. potatoes and sugar beet, instead of manioc and sugar cane). A different climate raises different problems as to soils, pastures, etc. Differences in the level of economic development result in the student having to deal with questions of management which are irrelevant in his own country.
— A stay of several years in a foreign country leads to the student losing his familiarity with the conditions of his own country. The student gets used to a standard of living — since the scholarships are usually quite generous — which stands in no relation to his future salary.
— The selection for scholarships of the most talented applicants works to the disadvantage of the domestic training programme. The average students lack the stimulus provided by the better ones. The sifting of talent in East Africa has already gone so far that the teachers there speak of the "local training system being ruined by this nuisance of scholarships".
— The capacity of local agricultural training centres is by no means exhausted. Hence, scholarships would not provide additional training opportunities, but would simply exchange inexpensive ones adjusted to local conditions for expensive ones under unaccustomed conditions.

The arguments against the scholarship system are directed against basic training given somewhere else, not against the further training of young scientists, after they have obtained their diploma. The experience of the extension service speaks in favour of a 6—10 month working stay abroad, during which the scholarship student would take part in the actual work, in order to gain experience in methods and intensity of work. The demonstration of various conditions in which production takes place would broaden his horizon, and the stay abroad would be short enough not to estrange the trainee to any great extent from the tasks awaiting him at home [1].

The observation that preference should be given to the furtherance of local training raises the question of which type of agricultural training would justify being supported in this way. In Tanganyika we can distinguish between the following kinds of training:

[1] The USA pay for agricultural officers from Tanganyika to take part for several months in extension work in the USA. The second stage of these scholarships consists in the student taking part in extension work at the successful American agricultural college at Saloniki, Greece.

Department of Agriculture, University College, MAKERERE, Uganda. The most important of the various forms of training at the moment is one preparing well-informed and thoughtful personnel for national leadership. 50 capable high ranking officers are certainly more important to the agricultural development of Tanganyika than 1,000 average Field Assistant Officers, as the work of the latter cannot really bear proper fruit if they are not given proper guidance. The training of people to be considered for executive positions should in the first place be provided at MAKERERE, followed by study abroad. In so far as development aid can assist in extending this kind of work at MAKERERE, it should be given preference.

Agricultural College, MOROGORO. The College is in the initial stages of being built up. Planning and management is done through US development aid, scientific competence and supervision is in the hands of West Virginia College. Students who have passed the 2-year course at the agricultural schools at TENGERU and UKIRIGURU, and those with a "Cambridge Certificate", will be given a 2-year advanced agricultural training at MOROGORO, terminating in a diploma. The training is chiefly, but not exclusively, a preparation for service in the Ministry of Agriculture.

The College at MOROGORO will probably constitute an agricultural centre for Tanganyika. Such a centre is needed. Morogoro is well situated with regard to transport facilities. All aid is useful which would contribute to the project starting off on a sound footing. A useful contribution might be to enlarge the scope of the institution, which hitherto has been under exclusively American influence, by other aid-giving countries making personnel available or establishing affiliated institutes. Such additions would also accord with the long-term interests of the Americans as the main trustees.

There is a chance that in the course of time the College may turn into a centre on the lines of the American "Land Grant Colleges", which would undertake extension services as well as training. If efforts are made in this direction, they would deserve careful support. The same applies to the association of the extension service with other research and training institutions in Tanganyika. It may well happen that the agricultural administration will become top-heavy, too much an instrument of administrative instructions and control: in that case one could not expect its extension service to be very effective. In future it might be better if agricultural extension work were concentrated on specific jobs and attached to research and training rather than to the administration.

Government Agricultural Schools. There are agricultural schools in TENGERU and UKIRIGURU where students can attend a 2-year training course, provided they have had 10—12 years of school education and worked for some time at district offices of the Ministry of Agriculture. In addition, there is the "Land Planning School" at MOROGORO for the training of land surveyors. All these schools are most necessary and would justify support.

Other Agricultural Schools. Problematic are the schools which intend to train actual farmers. An average farm consists of 2—7 acres of land with a gross yield of hardly more than £ 100 per year. The attempt to give future small-scale farmers lengthy training leads to a quite intolerable ratio of training expenditure to increased tax yield — due to greater production by those who went to the agricultural school. Also, it is not likely that after such training the trainees would want to go back to their villages and take up their work with the hoe or the ox-drawn plough again.

In view of this, there exist in the main three useful additions to the above-mentioned training facilities:

— Farming institutes, where the farmers can attend short courses. They are a favourite object of development aid.

— The training of "Rural Community Workers", with emphasis on subjects of an agricultural and social nature rather than on technical farming.

— An apprenticeship for farmers' sons on well-run farms, estates, state-farms, etc., supplemented by courses comparable to the German "Agricultural Winter School".

Every kind of agricultural training is of course useful in some respects. But as soon as training inputs and estimates about possible outputs and their tax returns are compared, institutions like Makerere, Morogoro, Tengeru, and Ukiriguru earn preferential treatment. The idea of giving the actual farmers formal school training cannot, for the time being, be accepted without reservations in Tanganyika. In view of the small size of the farms, the cost of training would hardly permit any other course than that training should be restricted to the instructors, i.e. extension workers, settlement officers, etc. Hence it follows that, at the beginning, it would be better to limit development aid to training institutions which are integrated in the national training programme. Otherwise, there is the danger that successful students might find it difficult to secure a worthwhile place in professional life.

Without denying that careful support for agricultural training is very important to Tanganyika, it must not be overlooked that there are not very many opportunities for development aid to be usefully absorbed. In 1962, the schools at Tengeru and Ukiriguru could not work to full capacity because there were not sufficient applicants with adequate education. It is not yet possible to make a selection from among the applicants and to restrict admission only to those who appear particularly suited. Agriculture has to compete with urban professions which as a rule seem more attractive. Boys with secondary school education realize that they are in great demand and exploit the situation accordingly. There is little devotion to the national cause. They think it is natural to get paid for education and they are difficult to discipline. The Ministry at Dar-es-Salaam dislikes the dismissal of unsuitable students because of the shortage of boys with sufficient schooling. Consequently, there is no point in building schools for which there are no students, or the students of which make excessive claims.

Also, the fact must not be overlooked that Tanganyika's capacity for raising taxes is small and that the financial outlay on training institutes must remain in reasonable relation to revenues. It is useless to establish schools and agricultural institutes which the recipient country cannot staff and run properly because of the lack of money and personnel. Aid in the field or in agricultural training therefore imposes certain obligations. It is not enough to give a sum for the purpose of providing the building. To what extent this aid is really useful depends on the job done inside the building. Hence any aid in this direction ought to include a careful check on the work done in these buildings and the willingness, where necessary, to help out with staff, in order to ensure that the investment will bear fruit.

There is at present no shortage of teachers in Tanganyika's agricultural schools. The process of rapid Africanization has brought about that many Europeans have been transferred from the extension service to agricultural schools. One cannot say at the moment whether this position will last. If a shortage of teachers were to occur, development aid could and should help to overcome it. Special knowledge of tropical agriculture is not needed. It is more necessary to have a good standard of theoretical knowledge and the ability to convey this in fluent English. A new arrival can share the job with the older teachers by teaching basic theory, while the latter would deal with special problems of tropical agriculture. It would appear quite practicable for young German agricultural teachers to spend five years at an agricultural school in Tanganyika as a step in their own professional careers.

3. Agricultural Extension

The British Administration was of the opinion — shared by the author — that a well managed agricultural extension service, supported by training, research, small loans and social work at village level, represents the most effective instrument for promoting agricultural development. This does not necessarily mean that it is a suitable field for development aid. Extension work consists of a great many single services, which change from place to place and from one year to another. This means pay and equipment for personnel whose activities it is difficult for the donor country to control, and the results of whose work cannot easily be assessed. This would suggest a division of responsibilities, i.e. development aid should take charge of research and training, while the recipient country should take care of introducing innovations by running the extension service.

Desirable as this division of responsibilities may be, it is no solution in Tanganyika. The expansion of agricultural extension planned first by the British Administration, and now by the government, is so extensive that in Tanganyika there is just not enough money for it, and still less qualified African personnel. The government of Tanganyika therefore stresses the need for development aid for the purpose of financing the budget of the Ministry of Agriculture, i.e. to pay for the general development services provided by the Ministry.

However, there are weighty arguments against such development aid:

— The donor country has little control over the use of such funds and the work done by the staff paid from them.
— A good relation between input and output in agricultural extension is a possibility depending on the guidance and supervision of the staff.
— The degree of usefulness in furthering peasant production depends to a great extent on puritan attitudes. Lack of care in the granting of loans and subsidies, subsidized prices and the unpaid-for supply of means of production, could lead to the peasants becoming accustomed to receiving "gifts", and hence to a lessening of economic activity.

In contrast to this, in large agricultural projects — estates, pilot farms, irrigation schemes, etc. — expenditure and likely returns can be approximately estimated in advance. Something visible, permanent and — this is the general idea — exemplary is brought into being: a nucleus of orderly land use. But such projects are costly. The money for just a few projects would exhaust the available funds. £ 1 million spent on irrigation could provide 1,500 settlers with about 5 acres each. With amounts of that order, some 100,000 peasants have been enabled to turn to the commercial production of cotton and pyrethrum, with the aid of extension and several administrative services. Large schemes, in the beginning at least, must be directed by non-African personnel. The population usually sees in them an expansion of the European estate economy. At best they might be welcomed as an added opportunity for employment, at worst — particularly as the people are not alive to economic inter-relationships — as the expression of a neo-colonial policy.

Hence, we can say that

— if a well-organized extension service is operating;
— and if there exist innovations which are simple, popular, and ready to be put into effect;
— if the extension work can be concentrated on promising districts, crops, and people;
— and, finally, if political and social conditions are conducive to the farmers taking up such innovations —

then — in view of the doubtful usefulness of large agricultural projects — preference should be given to the promotion of extension work. After all, the greater part of Tanganyika's agricultural export originates from African farms and, in the final analysis, is the result of ideas propagated by the agricultural administration and the extension service as the agents of technical progress.

In Tanganyika, however, the personnel of the Agricultural Department and its extension division have undergone rapid changes. There are officers devoted to their work. The peasants are interested. But what is needed now is consolidation, not expansion. This means that for the time being limits are set to the otherwise most efficacious use of agricultural development aid, i.e. the financing of the Ministry of Agriculture's expanding services.

A possible way out would be for the donor country to conduct with its own personnel certain specific extension schemes, while the recipient country agrees to transfer responsibility to foreign agriculturists. Such schemes might be considered on the following lines:

— A specific extension scheme is worked out, confined to a given area and concentrated on profitable innovations, similar to the successful "increased productivity schemes".
— The recipient country provides buildings, field assistant officers, community development workers, etc. It guarantees political support for the scheme.
— The donor country assumes the management of the scheme, pays the salaries, equipment, provides small loans, etc.
— The recipient country provides an experienced agricultural officer as adviser to the director of the scheme, who is unlikely to know the locality.
— The donor country selects a project, the priority of which is recognized by the recipient country.
— The size of the project may depend on circumstances. If this form of partnership proves successful, larger projects may be considered on the lines of the former Sukumaland Development Scheme.
— We would recommend regarding the first of these projects as a trial project, and financing it with a grant. If this proves successful, further projects should be financed on credit terms.

The suggestion that the donor country should take charge of the agricultural extension service in certain locations must not be mistaken for a plan to create model districts. To concentrate on one cash crop, or on certain progressive villages or single farmers, that is to say, on projects with a reasonable expectation of fiscal profitableness, seem feasible propositions. The setting up of model districts requires the promotion of all branches of agriculture, erosion control, etc. In Tanganyika, the time is not yet ripe for such an integrated undertaking; nor are conditions favourable for an overall land plan. The areas around Mt. Kilimanjaro and Mt. Meru are in need of a comprehensive re-organization of soil and water. From a technical economic point of view conditions are probably favourable for overall planning. But this requires the active co-operation of the population, or at least a ready acceptance of measures serving the common good. This readiness cannot yet be expected. The essence of the present suggestion is not that the personnel and money of the donor country should provide examples or models, but rather that certain especially profitable extension schemes should be carried out under the responsible direction of the country giving development aid.

4. Settlements and Production under "Close Supervision"

As against the limited opportunities for development aid in agricultural research, training and extension, the scope is enormous for settlement, production under "close supervision" and, in particular, for irrigation. The problems there are of marginal economies, partly because of low prices, partly on account of the human factor being an obstacle to the satisfactory running of suitable institutions.

When considering settlements it must not be overlooked that their function differs fundamentally from those in industrial countries, where usually social and political reasons preponderate. Since there the industrial activities are first in importance, it is perfectly feasible economically to finance from public funds certain structural adjustments in the relatively less important agriculture. In a development country like Tanganyika, rural settlements have a different purpose. Agriculture is the basis of the economy. It would be contradictory to subsidize it, as its role is precisely the contrary, i.e. to contribute to the funds necessary for the expansion of town and industry. If settlements mean subsidizing agricultural production, then such schemes are contrary to the demand for production increases at low cost. On the basis of such fundamental considerations we can classify projects of that nature in Tanganyika as follows:

(1) Land Reclamation

Unplanned Settlements. In various parts of the country farmers proceed to reclaim land in a haphazard manner (expansion areas). Although this merely represents an expansion of the customary methods of farming, it nevertheless leads to considerable production increases at little expense. This unorganized form of land reclamation is a type of settlement which accords with conditions in Tanganyika. It could be helped along by the construction of supply roads and the provision of drinking water (dams, wells). This kind of aid usually shows a high rate of fiscal profitableness, i.e. investments by the government show considerable additional fiscal returns after a time. Hence, the use of development aid for the construction of such roads and watering places may constitute a worthwhile proposition.

Planned Settlements on non-irrigated Land. In contrast to the above, planned settlements on non-irrigated land which are not subject to "close supervision" are not suited to development aid. The costs are considerable. The institutional prerequisites necessary to implement planning are lacking. We are not dealing with "alienated" land. In all probability conflicts of competence will arise with the local authorities as soon as attempts are made to carry out a rational organization of production.

(2) Producing under "Close Supervision"

Plantations. On the other hand, relatively good results are likely from the supervised production of crops with high gross returns per acre. This may be on settlements, as in the case of Virginia tobacco at URAMBO, or of sugar cane at SONJO (Kilombero), but could also involve the issue of farming licences, as in the case of peasant tea. In the beginning it will not be possible to recover the costs, but the long-term effects of introducing new crops in this manner could well be very significant and might justify the initial subvention. It is probable that aid-money will be well used as long as it concerns a high-yielding cash crop under "close supervision".

Ranching. The same is true of animal husbandry under "close supervision". It is no use giving development aid, for instance, in the form of mobile clinics, for the purpose of ensuring the health of cattle which are not sold because the owners do not wish to sell them. Apart from such destructive epidemies as rinderpest, the job keeping the cattle healthy is a useful task only within a rational system of animal husbandry.

The ranch and settlement at KONGWA represents a model incorporating, as a first phase, the establishment of a large-size development centre and, later, the association of neighbouring herdsmen with supervised cattle. Similar projects can be considered as forms of development aid with relatively good economic prospects, in addition to which they would promote a branch of agriculture which in future might become one of the most important in Tanganyika.

Reservations. A necessary condition for production under "close supervision" to be suitable for development aid is, in the first place, the enforcement of strict discipline and, in the second, that a sufficient number of staff personnel is available. The recipient country will probably be in no position to supply sufficiently qualified staff. The donor country, therefore, will have to provide both the money and the personnel and, at the same time, a project-bound training.

5. State Farms

In the course of giving development aid for the purpose of activating existing farms it might well be useful to establish some larger state farms. The term "Model" or "Demonstration Farms" is unfortunate, as it creates the impression that farms of this type could serve as examples for the surrounding peasants. As a rule, their influence as examples is small or even nil. The methods used on the large farms cannot be copied by the smaller ones: or it would be better to say that the idea is so widespread that what happens on the large farms has no bearing on the work of the African peasant, that no attempt is made to emulate them. Of all forms of example the large model farm situated among peasants is the least effective. In Tanganyika, as in other developing countries, the progressive neighbour, i.e. the example by way of agricultural extension, is incomparably more influential.

Nevertheless, the establishment of some state farms can serve a useful purpose:

— It is doubtful whether private estates will continue to make investments in innovations, try out new methods, and open up new markets. Political independence has slowed down these activities. It cannot yet be expected that medium-sized African farms will take over these functions in sufficient measure. Nor are independent farming co-operatives in a position to do so. In view of this, there is no other way but for the government to pursue such activities, possibly using the TAC or private firms as "Managing Agents".

— In some areas there are no estates at all, for instance in Sukumaland. They are, however, needed for providing improved seeds, trying out innovations, etc.

— Tanganyika offers relatively good conditions for ranching. Where private initiative does not suffice to exploit existing opportunities, or where political objections bar the extension of rights of occupancy to Europeans, it is appropriate for the government to establish further ranches from public funds and to put them in charge of the TAC.

— Although it may be a good thing to settle farmers on newly irrigated land, it is better not to proceed with this before the best use the land should be put to is known. It is probably desirable to run new irrigation projects for some years as large farms.

In any case, large state farms should not be an end in themselves but should be regarded as development centres for the purpose of raising the level of farming in the neighbourhood. Accordingly, two stages and two duties for large state farms may be distinguished:

(1) Developing and managing the farm, without being burdened with demonstration, training and research duties, but aiming at the greatest possible commercial success.

(2) An association of services for the development of the surrounding countryside. This might mean either production under "close supervision" or the establishment of a nucleus for extension, repairs, marketing, processing, etc.

Experience shows that almost everywhere in the world the running of large state farms is a costly business. This is true, too, of large farms run on development aid. In most cases, costs cannot be recovered. This need not necessarily lead to such projects being rejected. From an economic point of view their value lies in the direct or indirect influence they have on agriculture in the surrounding countryside. Nevertheless, it is advisable to keep costs low. The following principles should be observed:

- The costing should be realistic, i.e. better too high than too low. Within the framework of development aid all projects become matters of prestige. Once begun, it is hardly possible to drop them again. A revision of the estimates of costs results in delays in the costly initial phase of the build-up.
- Development centres or any large farms should be set up by the donor country only where there are particularly good opportunities for marketing, and favourable soil and climatic conditions. Taking over the management of an estate involves an obligation to set an example in the quality of crops and animals, and this may become costly indeed if the conditions for production are not very good. Special caution should be observed where land is to be cultivated which, up to then, has been lying idle. Almost always there are compelling reasons for this, which a stranger would not know. Only where it is known why the land has been kept idle, can an estimate be made as to the prospects of the project.
- The most practical size of such a farm will depend on circumstances. To open up large areas requires a lot of money. If only limited funds are available, it is better to farm small areas intensively and with high yields per acre than large ones with low yields per acre.
- Rights of tenure must be clearly established before the aiding country commences the work of construction. Under the existing conditions in Tanganyika the land in question must be "alienated". Under no circumstances should land be cultivated which comes under the jurisdiction of the local authorities.
- It is usual for this type of farm to be given too many duties too early. It would be better not to clutter up the initial phase of construction with training, demonstration or research duties.

6. Industries Processing Agricultural Raw Materials

Agricultural and non-agricultural development in practice go hand in hand with the very important task of the processing of agricultural raw materials. This is not the place to define possibilities and limits or advantages and difficulties of this form of industrialization. It suffices to point out that it is a most suitable field for development aid. Staffing difficulties are relatively slight. The preparation and direction of such enterprises can be transferred to private firms as managing agents. *Most valuable are those industries which obtain their supplies of raw materials from African farms.*

7. Development Aid and Irrigation

The great importance of irrigation makes it necessary to add to the observations on the most useful institutional basis for development aid — which in principle apply also to the various aspects of irrigation — some points relating specifically to this subject.

As far as agricultural development aid is concerned, we have to distinguish between help for small and help for large irrigation schemes.

(1) Smaller irrigation schemes are mainly the responsibility of the local agricultural administration and extension service. They organize construction, give technical guidance, and authorize subsidies or small loans. The input-output ratio is frequently not bad. Nevertheless, the promotion of such schemes through development aid needs careful thought. Small irrigation schemes can largely be installed and operated on the basis of self-help. A generous granting of subsidies and loans usually leads to the peasants holding back with their own efforts. Though some aid may be indispensable, it must not be overlooked that there are limits to the amount that can be employed usefully, if the peasants' initiative for self-help is not to be affected adversely.

(2) The situation is different when we come to consider *large irrigation schemes*. The scope is enormous. Development aid is able to create large and lasting achievements. The problem is that, with regard to large schemes, Tanganyika cannot yet be regarded as providing attractive opportunities, mainly for the following two reasons:

— Even with proper farming, the investment is generally not likely to pay off appropriate rates of interest.

— The mentality of the peasant population makes it unlikely that institutions could be established guaranteeing that the land is farmed properly.

Countries regarded as suitable for large-scale irrigation schemes are distinguished by the following features: either

— the products resulting from irrigation schemes are in great demand to supply domestic markets which have sufficient buying power;
or,

— the products can be sold at a very low price, either because of the great fertility of the land, the low costs of establishing irrigation, or low costs of transport;

or,

— crops can be planted with high gross monetary returns per acre. (for instance, sugar cane and cotton).

These conditions are in Tanganyika exceptional rather than the rule. Development aid for large-scale irrigation schemes would mean financing an investment which might pay hardly any interest at all and from which, at best, only a slow amortization could be expected.

Again, these are not necessarily arguments against development aid being used for large-scale irrigation schemes. Investments of this kind lay the foundation for economic structures which sooner or later have to be built anyway. This is not an argument against the idea of such irrigation schemes in Tanganyika per se, but rather against such investment at the present stage of development.

Against this, one could argue that to carry out large-scale irrigation schemes takes decades and that one should make a start now with projects which, in the course of time and changing economic conditions, will come to fruition when the situation is more profitable. In view of Tanganyika's geographic position one could consider a division of labour with India, where industrialization is proceeding at a rapid pace, and where agriculture offers less favourable conditions for production than East Africa.

In favour of investments in large-scale irrigation schemes is also the fact that other sectors of the economy are not able to absorb the volume of investments necessary for economic growth to take place. Of the industrialized nations, hardly any of them would have reached their present condition if they had invested exclusively in profitable undertakings. If there is no opportunity of following up projects with high profits, there is no alternative but to make do with less profitable ones, in this case irrigation.

The third, and perhaps decisive, argument in favour of an early start on large-scale irrigation schemes is that agricultural administration and extension need consolidating, not expanding.

In this situation it may well be sound policy to use development aid for the construction of large-scale irrigation schemes. Problems of staff in the Ministry of Agriculture, arising from Tanganyika's recent independence, may be considered a compelling reason to start large irrigation schemes now, although they would fit in better at a later stage of development. But it would be wrong to expect the recipient country to be able to pay interest and amortization on the agreed terms to the donor country advancing credits for this purpose.

III. A General Proposal Concerning the Technique of Selecting a Project

In the Federal Republic of Germany, the usual method of selecting projects for agricultural development is as follows: A group of experts tour the country for 3—4 weeks. During that time they receive suggestions for development aid which they assess critically, after which they submit their own recommendations to the Federal Government. The Federal Government examines, accepts or rejects these, or dispatches a further team of experts to examine the recommendations in detail or to prepare new ones.

This method functions satisfactorily if the developing country is in a position to submit plans for concrete and practical projects. The problem is that there is a shortage of information and experts to formulate a cohesive development policy, and a lack of technical experts to work out such projects. Hence the developing country is frequently not able to give succinct information on the projects, to recognize priorities, and to do the necessary preparatory work.

Quite often the experts themselves have to look for worthwhile projects and to recommend them for development aid. This clearly is beyond their scope. The most that can be achieved during a 3—4 week visit is to gather some ideas about a country's economic situation. But it is too short a time to come to conclusions on what should be done with development aid. Experts are very busy people. There is not enough time for them to become thoroughly familiar with the specific conditions obtaining in the country. As a rule they are not acquainted with the literature on the subject of what has been attempted already, and what has succeeded or failed. As these groups are composed, it is unavoidable that, for instance, an expert on co-operatives should have to give an opinion on irrigation schemes, and vice versa.

Agricultural development aid comprises a wide range of subjects, from problems of irrigation, animal husbandry, plantation crops, training, extension services, and co-operatives to questions of marketing organization and community work. In order to judge properly, it is necessary to understand the role of agriculture in the overall economic picture. In this respect, technicians — and agriculturists are technicians — tend to have strong and erroneous opinions. Agricultural economists or rural sociologists, on the other hand, tend to lack detailed knowledge of agricultural techniques. It is a mistake to assume that knowledge of the country alone qualifies. For instance, people active in estate farming will regard the promotion of agricultural extension in Tanganyika as a waste of time, whereas others, who have done useful work in that field, will scorn the idea of wasting money on irrigation schemes. In short, it is impossible during a 3—4 weeks stay to assess the relative advantages of different agricultural development measures.

The fact that the experts cannot cope can be seen from their reports. These contain either

— a cautiously worded general outline of the agricultural situation in the country concerned. The Federal Government does not obtain what is needed, namely proposals in project form, if possible with input-output estimates,

or

— precise proposals. Whether in fact these proposals accord with the actual needs of the country is not at all certain. The chances of wrong investments being made are great.

As a contribution to the discussion on methods of selecting projects we would propose the following procedure:

Aid for the Establishment of a Planning Commission. Developing countries are not always able to submit detailed and carefully balanced projects for development aid. In view of this, we would recommend that the developing country is encouraged to set up a planning authority and is given aid for this purpose both in money and personnel.

The Submission of Proposals. The Federal Government receives proposals for agricultural development aid from the planning authority of the developing country. These are examined in Germany by the responsible government officials as to whether they are basically suitable for German development aid.

Expert Survey. According to the nature of the project, experts are sent to the developing country to survey the proposed project on the spot. As to agriculture, there should always be two of them, one with specific experience in the techniques of the proposed project and another — preferably an agricultural economist — who is able to judge whether specific projects are appropriate to the general economic situation.

Government Decision. On the basis of the preliminary examination of the proposals submitted in writing by the developing country, and the subsequent report of the experts, the Federal Republic decides whether or not to accept the proposals, or, if need be, to ask the planning authority to submit alternative proposals.

IV. Summary

The observations on a purposeful implementation of agricultural development aid can be summed up as follows:

— *The most important form of agricultural development aid is the furtherance of urbanization and industrialization, i.e. the growth of urban purchasing power*. As far as capital aid is concerned, priority must therefore be given to sensible non-agricultural projects.

— A further important form of agricultural development aid is investment in *industries processing agricultural raw materials*. They create or improve market conditions and constitute, therefore, an excellent means to promote production.

— The priority given to industrial investments does not imply that agriculture is regarded as less important. On the contrary, it is argued that, in countries like Tanganyika, the important, and in many ways decisive, contribution agriculture must make to the growth of the whole economy, can be achieved by activating existing farms, i.e. without the state or an aid-giving country making priority investments in farms, equipment, irrigation, etc. It is rather a question of putting technical knowledge into wide-spread practice. *For this purpose development aid is indispensable in the field of applied research and training*. There is less need for new research and training centres than for the expansion of existing ones.

— A necessary complement to research and training is the spread of improvements and innovations through the agricultural administration and extension service. Dividing the labour between donor and recipient countries, these are jobs which should be tackled by the recipient country. If it is not in a position to do so, we would recommend *agricultural development aid for the extension service*. As in the case of research and training, this will be largely a matter of personnel.

— The introduction of technically ambitious production (e.g. plantation crops) into peasant farming requires that *production is kept under "close supervision"*. Whether aid should be given for such projects depends on whether the recipient country is willing to set up the necessary institutions and enforce their authority among the participating farmers. In cases where projects of this kind are found worthy of support, capital aid will not be enough: as a rule there will also have to be assistance with personnel.

In general, the above-mentioned proposals for development aid will not absorb the total volume of investment necessary to ensure continuous economic progress. With regard to agricultural administration and the extension service, the amounts that can be invested usefully are limited because of staff difficulties. Where this applies, it is advisable to confine aid to such projects as are relatively independent of the local administrations. In this connection we would recommend:

— The construction of *large-scale ranches as development centres*, to which — once they are working properly — promotion measures could be added for the benefit of the surrounding herdsmen.

— The organization of *large-scale irrigation schemes*. Because of bad market conditions, Tanganyika is not yet "ready" for such schemes, i.e. investments in them are not likely to yield reasonable interest or amortization. However, since other spheres have a limited capacity for absorbing aid, there may be no alternative than to invest in, as yet, unprofitable irrigation schemes.

Appendix A

Technical-Economic Approaches to Agricultural Development in Tanganyika

Our enquiries concerning agricultural development policy deal with the ways, institutions and measures, by means of which, the British and the independent Government were and are seeking to create production increases. As a complement we add the following observations on the possible direction production might take. Which crops to grow, or whether irrigation schemes and the use of tractors are worthwhile propositions or not, are questions requiring individual answers depending on soil and climate, the people concerned, and market conditions. Subject to investigation in each individual case, we can observe at the beginning of 1960 the following starting points for future agricultural development:

I. Cultivation of Crops

Sisal. The price of Sisal has risen from about £ 52 per ton to over £ 100 per ton, i.e. more than double. Present market conditions appear to favour the cultivation of more sisal, though it is, of course, difficult to forecast whether these conditions will last.

In addition, the sisal industry is facing a number of innovations which, if they prove themselves, might have far-reaching consequences:

Hybrid Sisal. Many years' work on breeding have resulted in a hybrid plant, which on trial plots has produced 10 tons of fibre per year and acre. On good soils and with mineral fertilizers the yields are four times as high as previously. In practice the fibre yield is expected to double. Good results were achieved, especially in comparatively dry areas. Another advantage of the hybrid is its longer growth cycle, which lasts approx. 15 instead of the usual 7—10 years, therefore reducing the annual acreage of new plantings. Hybrid sisal is already grown on numerous plantations. Not enough is yet known about its qualities. Plant diseases have occurred. A further obstacle is the destruction of the very much larger and harder trunk. If these difficulties can be overcome, the hybrid sisal would replace the present extensive cultivation by an intensive one. Fertilizing and plant protection would become economically feasible. The chances are that sisal production would be concentrated on favourable sites. The consequences would be shorter distances and lower transport costs. The obvious use vacated areas could be put to is ranching.

Weed Control. One of the most expensive jobs on a sisal estate is to keep grass and weeds down. The employment of chemical weed-killers might prove useful. Another method, which some plantations apply already, is the planting of perennial leguminosae in between the rows of sisal. These cover the ground and prevent the growth of grass and weed without affecting the growth of the sisal very much. In addition, they might have a fertilizing effect on account of their nitrogen-forming bacteria.

Animal Husbandry. Growing between the rows of sisal are grasses and leguminosae. This would suggest the keeping of cattle, particularly as grazing would reduce the cost of labour for weed control. Animal husbandry would, therefore, add to the estates' output while at the same time saving labour. The sisal-cattle combination is still at the trial stage. There are fears that the trampling of the animals in the rainy season will reduce yields. Furthermore, it has to be considered that cattle require veterinary supervision. Most of the estates are situated in tsetse-fly infested areas. That need not prevent successful ranching, but a careful check on health is essential. The size of herd to be kept on one estate, however, does not justify the employment of a sufficiently qualified cattle-man. Also, because investment in sisal itself is so promising, estate people are reluctant to spend a lot of money on building up cattle-herds.

Mechanization. The fact that wages have doubled within a year while prices are high has kindled a lively interest in labour saving equipment. It seems to be economic to mechanize.

Most important for hedge-sisal is an efficient decorticator. If present efforts to develop a decorticator which works without water are successful, i.e. if a good quality fibre can be obtained, hedge-sisal will gain significantly in competitiveness and might become a major industry.

Hence sisal is on the threshold of a technical advance, the effects of which can hardly be estimated yet in economic and social terms.

Tobacco. Few other crops need immediate development effort so much as tobacco, both flue-cured and Turkish. Soil and climate are suitable. Plenty of land which is marginal for other crops but good for tobacco is available, i.e. in the Southern Highlands, Western Region and Southern Region. It is doubtful, however, whether the principal tobacco-growing land up to now, around Iringa, is also the best. It would probably be better to expand production in areas where there are more favourable soil and climatic conditions. There is at present a good market for flue-cured tobacco from peasant farms. Some of the former Greek producers in the Iringa district have emigrated. The quota Tanganyika is entitled to within the East African Common Market is not being used up. It is high time for more tobacco projects under "close supervision" on the pattern of Urambo or Tumbi.

The beginning made with Turkish tobacco is also promising. The leaf produced around Tabora is said to be of good quality. Turkish tobacco is very well suited to hoe cultivation. The returns at 1,000 shs per acre are high and it is relatively simple to grow. In contrast to flue-cured tobacco, which should be grown under "close supervision", Turkish tobacco should be introduced into peasant farms with the help of the extension service.

Tanganyika's tobacco exports to countries outside the East African Common Market are at present confined to only small quantities. The most acute task is to fill the gap created by the departure of the Greeks. But one could well go further and regard tobacco as a prospective crop for export. For large stretches of the country it is relatively the best cash crop. Comparatively large incomes can be achieved from small plots of sandy soils easy to cultivate. It fits in well with hoe-farming. This applies, too, if the prices for producers are lower than present prices, which are above world market level. The export of tobacco requires, however, a sound market organization with sufficient financial backing to weather the costly initial years, und with enough power to induce the peasants and their co-operatives to sell a uniform quality leaf in regular quantities.

Cotton. With 30,000 tons of cotton exports, Tanganyika is one of the smaller cotton sellers on the world's market. Production has little influence on the price. There is no market quota. This would suggest that production should be encouraged. P. Bomani, Minister of Agriculture until 1961, set an ambitious target of 45,000 tons for 1966.

There is no lack of potential. The planting of cotton can be expanded in the Lake Region. The use of ox-ploughs, row-cultivation and weeders would help to increase acreage without reducing yields. There are large, relatively fertile, areas for more cotton in the Eastern Region and the Handeni District which can give high yields, provided insecticides are properly applied. In Sukumaland there are prospects of higher yields per acre. At approx. 400 lbs per acre, present yields are only about half of what could be achieved by introducing simple improvements, such as:

- correct spacing;
- early planting in November or December, i.e. during the "short rains". At the Ukirigura Research Station planting in early December produced 1,500 lbs, in mid-December 1,250 lbs, in mid-January 600 lbs and in early February 400 lbs seed of cotton per acre;
- early thinning and timely weeding;
- mineral fertilizing, if possible together with manure, which is at present left unused in the cattle bomas;
- tie-ridging in order to store water and to check erosion;

— complete picking. Up to now about 10—20 per cent of the crop was not picked. In particular, picking of the inferior cotton should not be neglected.

The cotton situation in Sukumaland is in many ways typical of the problems of agricultural development in Tanganyika. There is a chance to double output at little extra cost, if the extension service could convince the farmers of the benefit of the above-listed improvements. A repetition of the "Sukumaland Scheme" suggests itself as a successor to the scheme directed earlier by N. V. ROUNCE. On certain soils near the coast of Lake Victoria the startling effect of mineral fertilizing could help to introduce a whole series of improvements.

Another "Sukumaland Scheme" may, perhaps, be a way to offset the reduction of price subsidies in the summer of 1963. A farmer harvesting 50 per cent more is doing much better, even at 20 per cent lower prices.

Coffee. There is no lack of opportunity to produce more coffee in Tanganyika. The achievements of the coffee projects within the "increased productivity schemes" are proof that it is relatively easy to increase the acreage. Great opportunities exist to increase the yields per acre. On the slopes of Mt. Kilimanjaro, the best developed coffee area in Tanganyika, the farmers obtain on an average 3 cwt. per acre. Farmers in Kenya and estates in Tanganyika obtain 8—10 cwt. per acre. In the last four years the agricultural extension service has been trying to push up the yields by (1) the spread of better planting material, (2) mulching, (3) the application of insecticides and fungicides, (4) proper pruning of the trees, (5) uprooting old coffee and replacement by new plants, (6) replacing the mixture of banana clumps and coffee trees by planting in separate rows, (7) using Arabica coffee instead of Robusta coffee (where Robusta is the established variety). The estates have been able to raise yields significantly due to the joint effect of mineral fertilizers and copper spraying. The application of copper prevents the leaves falling off. With good foliage the plant can absorb more fertilizers. Efforts have been started to introduce this technical advance on the African farms.

Hence, the "technological leap forward" already a matter of fact in the case of estate-coffee is now imminent on the African farms. In view of the export quota for coffee, which is already being exceeded, this gives rise to some concern, the more so since it can be assumed that the trend towards greater coffee production will continue and can hardly be stopped. One can still earn good money from coffee relatively easily. It is in the nature of the stimuli provided by extension work that they will continue on their own volition to be effective, if proposed improvements and innovations have found ready acceptance in the first place, as happened in the case of coffee. The way to better incomes is, therefore, to produce a higher quality product. In this respect, the African farms in Kenya are well ahead in comparison with those in Tanganyika. The requisites for achieving higher qualities are mainly as follows: (1) careful husbanding of trees, (2) picking when the beans are really ripe and (3) processing the crop in central pulperies.

Tea. What is true for many other crops also applies to tea. There is no lack of potential. But the potential is to a great extent marginal with regard to rainfall, altitude, accessibility and available labour. In Kenya the situation is far more favourable. On the other hand, because of a shortage of other cash crops, tea is in many places a relatively profitable crop. The introduction of tea-cultivation on peasant farms deserves special attention. Some beginnings worth expanding have been made in the Usambara Mts., near Lupembe in the Southern Highland, in the vicinity of Bukoba, and near Tukuyu, not far away from Lake Nyasa. For these to succeed it is necessary to maintain "close supervision".

Pyrethrum. In Tanganyika the conditions for growing more pyrethrum are less favourable than in Kenya. The soils are poorer, the yields per acre smaller, and the pyrethrin content of the flower lower. On the other hand, many areas suitable for cultivating pyrethrum have no other cash crops with a comparable income potential. It is, therefore, in the interest of the peasants living there to produce more pyrethrum. This is true even in the case of lower prices. Up to the present the market situation is still quite favourable for Tanganyika. The processing firm has guaranteed sales up to 1967 of a maximum of 3,500 tons of dried flovers. With 1963 production at 2,700 tons, this leaves room for expansion. But because of the peasants' great interest in pyrethrum the quota apportioned will be fulfilled well before 1967. The question of further cultivation is, therefore, one of solidarity with Kenya. The price elasticity of demand for pyrethrum may be presumed to be small. Hence, greater supplies would have a negative effect on prices without any great increase in sales.

Cashew Nuts. Cashew nuts grow on good and on mediocre, on neutral and on acid soils. There is plenty of land suitable for cashew nuts in Southern, Eastern and Tanga Region. This is a simple crop. To establish the trees costs nothing but labour. Cashew nut production has spread with hardly any cost to the Treasury. In 1962 the prices fell a little. In view of a shortage of alternative sources of income and the small costs involved in planting trees, this should be no obstacle in the path of further expansion. In any case, the additional crops of the extensive new plantings of the past few years will have to be reckoned with in future.

Domestic processing might give impetus to further cashew production. At present the processing is done in India, apart from a test plant at Dar-es-Salaam. The Little Report on Tanganyika's industrialization gives the processing of cashew priority over all other industrial subjects.

Sugar Cane. In 1962 the sugar factory on the Great Ruaha River near Kilombero Valley started work. Run by a Dutch firm, it is financed mainly by British development aid. Sugar cane grown in the neighbouring Sonjo-Settlement under the "close supervision" of the TAC is also to be processed here. The target in 1962 was a production of 13,000 tons and, from 1964, of 35,000 tons. Plans are being considered to increase the capacity of the plant.

In 1963 sugar production in Tanganyika will probably reach 55,000 tons, thus making the country almost self-sufficient. The income elasticity of demand, however, is probably high. The report of the World Bank estimates consumption in 1969 at 73,000 tons. There is certainly room for more internal production. Suitable land with a good climate is available. Projects are being considered, or are in the process of realization, in the Lake Region and near Bukoba. It would probably be helpful to try and secure more supplies of sugar cane than at present from African peasant farms, using the estate as a nucleus for the development of surrounding areas.

Wheat. Successes in breeding rust-resisting varieties have helped wheat production to its sudden rise. The harvest in 1962, with over 16,000 tons, was significantly higher than that of previous years. Mechanized African farms are taking a growing share in production. Their production in the Northern Region amounted to 1,400 tons. There is perhaps more potential for wheat in Mbulu District, some parts of the Masai Reserve and the Southern Highlands. If this proves correct, the cultivation of wheat could be given a worthwhile impetus by measures to assist the maintenance of tractors and combines, and by mediation in the conflict between herdsmen and wheat farmers who compete for the land. Mechanized wheat farming is more popular than most other forms of farming. The demand situation is comparatively good. Domestic production could replace imports. Consumption is increasing at an annual rate of 7—8 per cent.

Rice. There are many reasons why irrigated rice is also a most valuable crop in Tanganyika — (1) it gives high yields per acre, (2) checks erosion, (3) thrives on different locations, (4) as a monoculture, simplifies the work of the extension service, (5) is open to significant increases in yield per acre through plant breeding and fertilizing, (6) can easily be stored, (7) is a popular food, (8) probably has a high income elasticity of demand and (9) is imported at present.

One can probably reckon on consumption increasing rapidly. Higher incomes have the effect of turning people from manioc and sweet potatoes, via bananas, millet, sorghum and maize, to rice and wheat. There are vast opportunities for growing rice in small irrigation schemes. The promotion of rice irrigation could largely be brought about through local self-help in the form of co-operative efforts. It would appear desirable to meet the domestic demand for rice from peasant production. Growing rice on larger subsidized irrigation schemes for the inland market could result in nipping the striking countrywide interest in irrigated rice in the bud.

Manioc. This unassuming but capable plant, growing on bad soils and producing high yields and fair incomes on good ones, could have an importance comparable to that of the potato in Central Europe. The peasants have got a great potential in this crop, which still awaits commercial exploitation. There is only a small domestic demand. Dried manioc amounting to 20,000 tons is exported annually.

Production has been given a lift by the breeding of a mosaic-disease resisting variety. On good soils the sale of dried manioc can realize up to 700 shs per acre (within 18 months). This justifies considering manioc as a cash crop. With no other crop can production increases be achieved by such simple and safe methods.

Cocoa. Some areas of Tanganyika are likely to have a certain cocoa potential (eastern foothills of the Usambara Mts., Kilombero Valley, south of Morogoro and at Lake Nyasa). Seedlings have been sold in Tanga Region since 1958. In 1963 there were 189 farmers with cocoa, having planted 20,000 seedlings. As to soil and climate, a small but promising area is available close to Lake Nyasa. 15,000 seedlings have been sold there up to 1963. Cocoa will hardly become a major cash crop in Tanganyika. It will be a valuable addition in certain locations.

Kenaf. In the East African market there is room for more sack manufacture. A possible fibre plant is kenaf, a hibiscus variety. There is a comparatively good market for it. One factory in Kenya needs 4,000 tons of hibiscus fibre. Attempts at cultivating kenaf in the Kilombero Valley appear hopeful, though yields are still rather low. There is probably scope for efforts in plant breeding.

Kenaf is probably suitable for peasant cultivation. It is of paramount importance for the growers to adhere to the correct harvesting time of 10—14 days. The cut stalks can be put together and left to dry in the fields. The separation of stalk and bark can be carried out by a power-driven decorticator which is brought to the field. The dried bark is what the farmers would have to sell.

Castor Seed. Up to now the production of castor seed in Tanganyika is based less on regular cultivation than on collection from wild plants. The plant is well adapted to the natural conditions of this dry country. The prices, at 40—55 cts per lb, are comparatively good.

The problems are low yields per acre and high costs of harvesting or collecting the seed. Attempts have not proved successful to grow a plant justifying the use of tractors and combines. On the other hand, castor-seed is regularly grown in considerable quantities in South Africa. It may be that the plant merits more money being spent on research.

Coconuts. Considerably more and better coconut palms than at present could grow along the coast of Tanganyika. It would be necessary to plant more, to protect the saplings from bush fires and to dry the nuts properly. Negligence and lack of interest are here, more even than elsewhere, the reason for the small volume of production.

Groundnuts. The groundnut scheme collapsed, not because soil and climate were unsuitable, but because it did not pay to engage in mechanized farming. Groundnuts are a worthwhile crop for producers using the hoe or ox-plough. It is probably more profitable to grow cotton, pyrethrum, coffee, rice, tobacco etc., but these cannot be grown everywhere. In some places groundnuts are the best choice as a cash crop. The market is regarded as relatively secure, particularly as a growing domestic consumption is observable.

Onions. Tanganyika is importing onions from Syria and the Lebanon. To replace these imports by domestic production would seem an obvious aim. Onions grow excellently in Tanganyika. Densely populated mountain areas are best suited for the cultivation of onions.

Bananas. There is a growing domestic market for bananas. In some locations they become a cash crop. The introduction of the Jamaica variety on suitable sites results in higher yields and better qualities.

Rubber. The TAC is trying to reintroduce rubber growing in the Kilombero Valley.

Oil Palms. Along Lake Tanganyika there is a small area where oil palms grow. Other areas may well be suitable. Nurseries have been established at Lake Nyasa. With irrigation, oil palms are perhaps a suitable crop along the lower reaches af the Pangani, Wami, Ruru and Rufigi. Market conditions are thought to be good.

Sesame. New varieties of sesame are producing higher yields. Prices are good. A trend towards increases in production is noticeable. As a cash crop, sesame is relatively simple to include in the traditional mixed farming of the Eastern Region.

Soya. Soya, too, is becoming a possible cash crop, thanks to progress in plant breeding. The aim is to grow maize for subsistence, interplanted with soya as a cash crop. New soya bean varieties planted between maize yielded about 400 lbs per acre in field experiments, without any adverse effects on the maize yield. Good conditions for the cultivation of soya can be expected mainly in the Eastern Region and the Handeni District.

Seed Pulses. For some years now African farms have been growing seed peas and seed beans under contract. An expansion of this rather profitable cash crop depends largely on the market being systematically organized. The aim should be to obtain more peasant production at reasonable prices in close co-operation with the seed firms, which buy the product.

Maize. Varieties with higher yields have been bred for some areas and are being distributed.

II. Animal Husbandry

Ranching. Since there are wide areas where rainfall is uncertain, animal husbandry in the form of ranching could become one of the most important industries of Tanganyika. It is possible nowdays to keep the usual cattle diseases under control. Large areas are free of the tsetse-fly. Selective bush clearing could rid others of it. Experiences at the Amboni ranch on the coast south of Tanga have shown that with very strict supervision the loss of cattle can be reduced to an economically bearable degree, even in the middle of tsetse infested areas. A grave obstacle to ranching is the high cost of bush clearing, and water supply. Among the vast stretches of land in Tanganyika there are, however, probably quite a few areas where ranching could be started without much bush clearing and where the supply of water could be secured at little cost.

It will take time for African cattle to be rationally organized. The necessary cultural changes can hardly be pushed through by force. The first step towards more rational animal husbandry would be the stocking up of existing TAC farms at Kongwa, Ruvu and Mkata. This could be followed — if the economics of ranching prove to be successful — by the establishment of more ranches. According to the calculations contained in the report of the World Bank, proper ranch management can be a commercial success. If the attempts at Kongwa and Matengoro of African cattle under "close supervision" continue to be successful, TAC ranches could in time turn into development centres for the improvement of African cattle in surrounding areas. The rationalization of African animal husbandry deserves particular attention, because the chief production factors — cattle, grazing, water and labour (for bush clearing) — are already given. What is required is not so much additional investment, but a better use of the existing resources.

Before further ranching projects are considered, the question of markets must first be cleared. Better quality meat is coming on the market. The meat processing industry is out to buy up scrap animals produced with todays' herding practices. The internal purchasing power can hardly absorb the quality meat produced on existing ranches. The export of refrigerated meat is not as yet feasible. Potential import countries bar refrigerated meat from Tanganyika because it is feared that the diseases occurring occasionally — like rinderpest — might be passed on.

Milk and Grass. Tanganyika has to import a fair amount of milk products. On the other hand, it has a substantial potential for dairy farming. High prices of 3—5 shs/gallon of milk are paid, particularly where some purchasing power exists thanks to a dominating cash crop like sisal, coffee or cotton. Successful beginnings in African dairy farming have been made by the Wachagga on Mt. Kilimanjaro where more and more exotic cows are kept. Hence, grass becomes an additional or alternative cash crop to coffee. It is an addition when, in the course of terracing, grass is sown to strengthen the terraces: it is an alternative when sown in fields. It has been suggested that grass fed to milk cows can produce higher gross yields per acre than coffee or bananas. Special attention should be given to fodder grass as a cash crop on higher grounds which are unsuitable for coffee. Hence, fodder grass varieties (setaria splendida, leucaena glauca, Rhodes grass and, on lower or drier ground, cenchrus ciliaris, elephant grass) are crops of possible economic importance.

Profound changes are necessary to make dairy farming a paying proposition — the growing, harvesting and storing of fodder, veterinary supervision, obtaining one calf per cow and year, careful rearing of calves, selection of good breeding bulls etc. All of this lies outside the sphere of local agricultural experience. To introduce such changes it is necessary to have strong price inducements or to direct production under "close supervision" [1]. In the beginning, financial assistance for dairy farming by subsidies for dairies, transport and the cost of personnel will be hardly avoidable. Since the introduction of proper dairy farming is a step of great importance on a long-term basis, aid for these purposes is more appropriate than for others.

Wool Sheep. In the Southern Highlands there are wide open areas which might prove suitable for wool sheep. Quick results cannot be expected. An initial stage of trial and error will have to be passed. It is apparently possible to keep the animals free of worm diseases i.e. to keep them in good health. The likely input-output ratio, however, has not yet been established. It is still too early for the peasants to engage in Keeping exotic sheep. It would be better if, for the time being, some larger experimental farms were to gain experience locally. These estates could be regarded as development centres for subsequent sheep-keeping on African farms.

III. Irrigation Schemes

There are differences of opinion as to whether large irrigation schemes would provide a sensible starting point for present agricultural development in Tanganyika. According to FAO estimates, about 74 Milliard cbm of water per year flow unused into the Indian Ocean. That is more than the water Egypt gets from the Nile. In consequence there is more than enough water for irrigation purposes, and there is also no shortage of land which could be irrigated.

---

[1] As an example of dairy farming and pig-raising under "close supervision", we would refer to the Sakay on Madagascar. See DUMONT, R.: Evolution des Campagnes Malgaches, Tananarive.

a) Areas with Great Water Potential

Rufiji Basin. The basin of the Rufiji river and its great tributaries, the Kilombero and Ruaha, comprises almost one fifth of the whole country. Of roughly 3 million acres of arable land spread about these three rivers, about 1.8 million could be irrigated. In addition, there are the advantages of flood control of non-irrigated areas. The costs of opening up the Rufiji basin through irrigation are estimated by the FAO at £ 170 million (see Table 32).

Table 32. *FAO Plans for Irrigation in the Rufiji Basin*

Location	Net irrigated area in 1000 acres	Capital costs in 1000 £	Gross Annual Output in £-moderate close		Proportion of Production to Costs-moderate close	
			supervision			
1st Stage						
Usangu	69	4,675	1,372	2,059	29	44
Kilombero	11	1,032	225	337	22	33
2nd Stage						
Usangu	133	12,383	2,669	4,004	22	33
Kilombero	107	14,027	2,146	3,219	15	23
3rd Stage						
Usangu	466	44,669	9,320	13,980	21	31
Pawaga	12	1,000	240	360	24	36
Kilombero	824	72,001	16,480	24,720	23	35
Lower Rufiji	200	19,030	4,000	6,000	22	33
Total	1,822	168,817	36,452	54,679	22	33

Source: Rufiji Basin Survey, FAO, Vol. IV, Tables 30 and 31.

Ruvu Basin. The Ruvu river has its source in the Uluguru mountains and flows in a broad valley through the country behind Dar-es-Salaam to the coast. Irrigation works could protect about 430,000 acres against floods and irrigate 290,000 acres. According to the Water Development Report, costs are estimated at £ 90 million.

Pangani Basin. The Pangani river is fed mainly from the mountains of the Northern and Tanga Regions. Its course runs through relatively densely populated country. A dam is under construction south of Moshi, primarily as a source of power. The land below this is regarded as little suitable for irrigation purposes, even though larger pockets of suitable land are said to exist.

Wami Basin. The Wami basin has its source north of Kilosa in the Nguru mountains and flows into the Indian Ocean about 46 miles north of the Ruvu. The amount of water it carries is roughly equal to that of the Ruvu. The costs of engineering works are assumed to be lower.

Wembere Steppe. Prior to 1914 the German Colonial Administration had plans to irrigate the Wembere steppe south of Lake Victoria. It is below the level of the lake, and should be fed from the lake by canals. The irrigatable land comprises about 500.000 acres.

The Shore of Lake Victoria. Numerous smaller rivers flow into Lake Victoria and offer scope for irrigation. There is also the possibility of shore-irrigation through pumping stations and sprinklers.

Ruvuma Basin. Hardly anything is known about the water and soil conditions around the Ruvuma, which separates Tanganyika from Mozambique.

The FAO examined the following areas in 1963 with regard to their suitability for irrigation: the *Plain of Rungwe* near Lake Nyasa, the *Luiche Delta* on Lake Tanganyika and the *Bubu Basin* in the Central Region.

b) Input-Output Ratio of Irrigation

The main problem of irrigation schemes in Tanganyika is the unfavourable input-output ratio. The prices for agricultural products are low and the internal market is small. Consequently investments in irrigation schemes aim to increase the export of agricultural products. Most of the areas listed above, moreover, would involve considerable transport costs from the project down to the coast.

No detailed investigations are available of the economic aspects of irrigation, except for one as yet unpublished study about the water usage of the Pangani river. According to rough calculation, the FAO reckons with gross investments of £90 per acre in the Rufiji Basin and annual gross returns, depending on whether there is "close supervision" or not, of between £20 and 30 per acre, i.e. 22—33 per cent of the gross investment. This corresponds to a ratio between gross investment and annual returns of 5 : 1 or 3 : 1. As a guide which, of course, needs checking in each individual case, it is assumed in the World Bank Report that this ratio must be 3 : 1 or below, in order to ensure for the Treasury that investments in irrigation bear interest at $8^3/_4$ per cent. According to the FAO calculations, therefore, large irrigation schemes must be considered marginal from the point of view of fiscal profitableness.

Moreover, the FAO estimates are rather optimistic. The World Bank Report contains estimates of £100 to £150 gross investment per acre, with annual returns of £20 to £30. This would represent gross investment to output ratios of 3.5 : 1 to 7 : 1. The World Bank states that interest rates of $8^3/_4$ per cent are attainable only if the annual gross returns per acre do not fall below £35. Of this amount, £7 per year and acre are set aside for amortization and interest and £8 for expenses. This would leave £20 per acre for the settler, representing a family income of £100 from a holding of 5 acres, which is about the size one family can manage.

This calculation too, is probably too optimistic.

— GROENVELD has tabulated the gross investment costs of irrigation projects estimated by the World Bank [1]. They amount to approx. £220 per acre, including the building of dams and a network of canals. The gross investment in irrigation schemes, in which the construction of dams serves more than one purpose, and in which only the main canals and the costs of land reclamation are debited to agriculture, amounts to £110 to £140 per acre. There is no reason to suppose that irrigation schemes would turn out cheaper in Tanganyika than elsewhere. On the other hand, it is unlikely that an annual average of £35 per acre can be achieved. The estimates for the trial project at Mbarali are £18 per acre. The actual gross return during the first two years fell well below that. Cash crops with high monetary returns per acre, such as sugar cane and cotton, can be grown to only a limited extent. Most of the areas with an irrigation potential lie in a belt in which no cotton should be grown, in order to prevent the spread of insects prevalent in Mozambique. The growing of sugar cane is restricted to the demands of the inland market. This leaves only one main cash crop, which gives — in other parts of the world — sufficiently high gross returns: rice. In view of the low internal purchasing power, however, it is likely that the price for rice will fall significantly as soon as greater supplies reach the domestic market. Hence, larger irrigation schemes will have to rely on the export of rice. This means that Tanganyika would have to compete with countries where the costs of producing and exporting rice are very much lower.

— A family income of £100 will hardly be enough to attract active settlers. Most of the project areas are sparsely populated. It would, therefore, be necessary for people to move in from overpopulated areas. It is doubtful, if not improbable, that a family income of £100 is enough incentive for good farmers. It must be considered that active individuals — the ones needed for successful settlement — usually have an income which is already above the average. Economically, the best way would be to use hired labour in order to farm newly irrigated land. This, however, is contrary to the intentions of the present government.

— Finally it is a question whether, in the first years of independence, the country is in a position to impose on settlers appropriate repayments of investments in water and soil as well as production under "close supervision".

To sum up, it can be said that it is most unlikely that an appropriate rate of interest can be achieved on irrigation investments, even with proper technical and economic management. Because of the human factors involved

[1] GROENVELD, D.: Investment for Food. Amsterdam, 1961, p. 132.

it may well be, that apart from a few particularly favourable projects, such as the growing of more sugar cane, no interest at all can be paid, and that even amortization will take half a century.

IV. Ox-Ploughs and Motorization

Under present conditions in Tanganyika it is not necessary to give priority everywhere to a replacement of the hoe by the use of oxen or tractors. Considerable production increases could be achieved by better husbandry on the existing acreage and by concentrating on crops with high gross yields per acre. But though there are many such opportunities, there can be no doubt that in the long run it will be necessary in most places to use oxen and/or tractors.

a) The Ox-Plough

The practice of using oxen and donkeys as draught animals has become popular in recent years, particularly in the Lake Region and the Southern Highlands. It is not easy to explain why for such a long time people did not show any interest in oxen as draught animals, and why they have suddenly come into fashion. After all, already 60 years ago, the German Administration tried to spread the idea of ox-drawn ploughs. In some pockets in Tanganyika, and in the neighbouring African countries, they have been in use for a long time. In some places, the ordinances of the British Administration in the 1920's and 1930's may have prevented the spread of ox-ploughs [1]. This is, however, certainly not a sufficient explanation for the interest in ox-ploughs coming so late. The sudden interest in ox-ploughs is probably a consequence of commercialization through the increased cultivation of cash crops. More land is tilled in order to earn more money, and that is labourious. The saving of labour through the plough, as against the hoe, is a great attraction.

On the other hand, displacement of the hoe by the plough is not without its problems. Mention of the important difficulties points at the same time to the forms of development effort which might bring improvements.

— Oxen as draught animals are not at all a cheap proposition for some farmers. To a remarkable extent, cattle is owned by only a few families. For many farmers who would like to use oxen it would mean having to acquire them first. Ploughing almost everywhere requires four oxen. This represents an investment of 1000 shs, plus 110 shs for the plough. Relying on these figures, COLLINSON has made estimates of the relative costs of alternative forms of cultivation at Usmao, Sukumaland. If

[1] The League of Nation's Mandate stated that the country's resources had to be protected and maintained. The change from hoe to plough increases erosion and in some places was, therefore, regarded as being contrary to the task given to the British under the Mandate.

cultivation costs by hired labour are put at 100, the following comparisons are obtained:

Hired labour (3/6 d. per day in cash and goods)	100
Ox-plough	117
Tractor (hired)	211

In this case ox-ploughing is more expensive than paid labour with the hoe, the quality of which is usually higher as well. Also, it must be borne in mind that COLLINSON is basing his calculations on fairly high wages. In general the farmers can draw on labour from their kinsfolk and friends, who are unable or unwilling to find a regular job for themselves, and who do not mind working for less money. In these circumstances, for the man who has to buy the beasts, ox-ploughing becomes an interesting proposition only if no labour is available during the peak months. Strong incentives are necessary to overcome the obstacles of high initial investments. Such incentives can be found in the offer of preferential loans. As long as repayment can be arranged in a technically simple way through the sales monopoly of a co-operative for instance, it is a useful and effective measure.

— The buyer of draught oxen takes a considerable risk with regard to the health of the animals. Frequently the beasts are not immune to local East Coast fever. Dips are absent, or the people don't use them. The health risk is especially great in large areas visited occasionally by tsetse flies. Hence a condition for ox ploughs to have a fair chance of success is easy access to veterinary help, or the buyer of oxen must be trained in methods to keep them healthy.

— Animal traction requires the land to be cleared of tree trunks and roots, and fallow grassland has to take the place of fallow bushland. Furthermore, larger fields are required. It is no longer possible to pick the best plots for cultivation, as one can when using the hoe, termite hills, for instance. Laying out larger fields facilitates erosion. Where the plough enters areas previously cultivated with the hoe in the form of five-foot ridges, it leads to a grievous deterioration in the quality of cultivation. The establishment of hoe-built five-foot ridges means a thorough mixing of the soil, a careful check on weeds, and offers first-class protection against erosion. The introduction of ox-ploughs usually implies changing over to the technically inferior flat cultivation. The smaller three-foot ridges, which can be established with ox-ploughs, are incapable of holding the water of the stronger rains.

This is not to imply that farmers should stick to the hoe. The point is that one should not ignore the danger of ploughs. There should be no ploughing up and down the slope. Erosion damage after rain should be repaired immediately. Some administrative ordinances in this direction are probably unavoidable.

- A cardinal problem is weed control. In Sukumaland, in places where ox-ploughs are used, the yields per acre have been reduced, sometimes to such an extent that, despite the fact that more land is planted, the yields per farmer are not greater than before, when the hoe was used on less land. Ridging necessitates planting in rows. Consequently weeding with the hoe can be done without much difficulty. Ploughing, however, leads to the technically retrogressive broadcasting of cotton seed. The farmers do not know the principle of planting in rows, unless ridging dictates it. In consequence, (1) more acres have to be weeded where (2) weeds can grow more freely, since the plough cannot do as thorough a job as the hoe, and (3) weeding becomes more bothersome because broadcasting has replaced the ridge-bound rows.

  Hence the use of ox-ploughs requires the introduction of planting in rows. As soon as that is done, additional implements like ridgers and ox-drawn weeders can be introduced to great advantage. This is mainly a task for the extension service.
- At present draught animals are used almost exclusively for work with the plough. Another important item would be the ox cart and, through it, the transport and application of manure. As long as the transport of manure is confined to head-baskets, there are no prospects of proper fertilizing. And without it, agriculture will remain soil-mining.

  The introduction of the ox cart has actually begun. It is frustrated on the one hand by the high prices asked for the carts and, on the other, by lack of roads. There are footpaths in a country with hoe-culture, but no cartways. The situation could be improved by subsidizing the production of carts. The construction of cartways is more suitable than many other measure for collective efforts based on local self-help.
- Draught oxen — like other cattle — feed on grass. After the dry season they are weak and not capable of much work. Consequently farming with oxen requires proper feeding and the growing and storing of fodder. It will take a long time for these practices to become general. A start could be made with some individual farmers who are interested. It is probably easier to induce the farmers to feed their draught animals properly than their other cattle. The feeding of draught animals could perhaps pioneer the introduction of feeding as a general principle of animal husbandry.
- The plough works more land and reduces the grazing land available. Rich families, whose wealth depends on their owning many head of cattle, will naturally regard the use of the plough as a danger to the basis of their social standing, their property in livestock. The present system of land tenure, according to which land under cultivation is private, while grassland is common and open to all cattle, forces the owners of large herds — and they are usually the local ruling group — to adopt a

hostile attitude towards the plough. There is yet another consequence. Farmers using a plough will invest their profits chiefly in more cattle. Less grazing due to ploughing, plus more livestock through purchases, will result in overgrazing and erosion damage.

Hence, the rational use of oxen also requires changes in land tenure and, in particular, a rational method of organizing the use of grassland.

b) Motorization

The facts that

— tractor use in connection with the groundnut scheme and the estates of the TAC has proved excessively costly,

— only about half of the cost could be recovered in attempts, in the early 1950's, to provide contract ploughing through Local Authorities and the Ministry of Agriculture,

— a relatively large number of the loans made to individual farmers for the purchase of tractors was not paid back,

— the costs of farming with tractors are higher than hired labour using the hoe, or the use of draught oxen (according to COLLINSON's estimates for Usmao, Sukumaland, they are twice as high),

— cultivation with ploughs is technically inferior to traditional practices with the hoe, hence erosion must be expected to spread considerably, at least in the initial stages of motorization,

all this is no sufficient reason to restrict development efforts to just the hoe or animal traction. The use of tractors has a "symbolic" and economically strategic significance. Animal traction can bring advantages, but it does not create much enthusiasm. What is lacking is a visible break with the way of living of the past. More than practical reasons are necessary to overcome such difficulties as the absence of cartways, the clearing of treestumps, the introduction of planting in rows, etc. The peasants not only take up the measures from which they can expect to derive a benefit, but also, and especially, those arousing their enthusiasm. In Tanganyika they are not yet so much under pressure from taxes or agitated by the desire for more goods as to be compelled to weigh up carefully the advantages and disadvantages of various forms of cultivation. The tractor is a symbol, and there is a readiness to invest savings in tractors which would otherwise be spent on consumer goods or the purchase of cattle. There has been a slow but steady increase in the number of Africans, Arabs and Indians who own tractors and hire them out to others [1].

---

[1] The "African Development Fund", started with American aid, advances loans for the purchase of tractors. Prospective buyers must raise a quarter of the price: to this end several related families often combine. A member of the group is usually put in charge who has been a tractor driver on one of the estates and has there learnt how to take care of one.

It must also be pointed out that input-output ratios do not give a complete picture of the advantages and disadvantages from a national point of view. Other economic and industrial activities are stimulated by the need for spares, repairs, better roadways, additional implements, trained tractor drivers and mechanics. The use of the tractor is of key importance in a sequence of economic relationships. Moreover, the tractor can be regarded as pioneering an active attitude of mind among the farmers. Those working with it are becoming more integrated in the division of labour and, at the same time, are obliged to think commercially. The work capacity of draught animals is lower in the tropics than in more temperate zones. Hence the tractor can substitute for a greater number of animals there. Eventually the tractor will establish itself in any case. Therefore it may well be justified to subsidize — although with caution — an unprofitable beginning in this direction.

The question remains who should be encouraged to own tractors. Under the existing marginal conditions of motorization, it is doubtful whether governmental, co-operative or communal tractor stations would have much chance of success. Experience almost everywhere in the world shows that this becomes excessively expensive. The best way would probably be the granting of preferential loans to those able to make a down payment of 10—25 per cent, and who are willing to offer their property (e.g. cattle) as security. In this way a considerable amount of savings among the rural population can be directed towards investment in tractors and thus towards productive ends. We would therefore recommend a continuation of the present policy of advancing loans facilitating the purchase of tractors, even at the risk of a considerable portion of these loans not being repaid in a regular and orderly manner. A further important measure of promotion is the establishment of a network of tractor workshops.

Appendix B

Marketing Boards and Market Controls

1. Marketing Boards

In Tanganyika there are marketing boards for the chief agricultural export products. Their members are appointed by the government. The various boards have different functions:

The *Lint & Seed Marketing Board* sells Tanganyika cotton (lint) abroad and cotton seed to domestic and foreign oil-mills. It supervises the work of the ginneries and controls the quality grading of cotton. The Board made considerable gains in the years of the Korea boom. In 1955 they amounted altogether to £4.7 million. From its current and accumulated income the Board finances, by way of loans and grants, projects promoting the cultivation of cotton in Tanganyika and the establishment of co-operatives. Since 1955/56 the reserves have been mainly used for price subsidies.

The *Coffee Board* had only a consultative function up to 1962. It financed research projects and advised the government in respect of the International Coffee Agreement. Since July 1962 the board has partly taken over the sale of coffee supplied by the co-operatives. Previously this was done entirely by the coffee co-operatives themselves.

The *Pyrethrum Board* acts as a mediator between the producers and processing firms, negotiating 5-year contracts on the price of the dried flowers. It advises the government on the amount of land which the terms of these contracts would justify cultivating. The government, by the issue of licences, keeps a control on the area to be cultivated in accordance with the market estimates.

It is planned to set up a *Tobacco Board.* The output is sold mainly in East Africa. Annual contracts with the producers and guaranteed prices for certain qualities have contributed to the stabilization of prices.

The trade in sisal is subject to no controls. *The Tanganyika Sisal Growers' Marketing Organization* concerns itself with internationally recognized qualities in a smoothly working sales system.

An Agricultural Products Board started work in 1963. It is to concern itself mainly with those products the export of which is not in the hands of any of the above-mentioned boards.

Table 33. *Agricultural Market Organizations in Tanganyika*

Crop	*Commodity Board or Organization*	*Principal Functions*
Sisal	Tanganyika Sisal Growers' Association	Producers' organization for consultation on matters of common interest and for the promotion of research.
	Sisal Marketing Organization	Sells sisal on behalf of its members, supervises grading.
	Lake Region Hedge Sisal Board	Organizes sales of hedge sisal.
Cotton	Lint & Seed Marketing Board	Organizes transport and sale of cotton and seed, supervises grading, advises the government on price policy.
	Cotton Board	Advises the Agricultural Department on questions concerning cotton.
	Co-operatives	Purchase the crop, transport it to the cotton ginneries, operate some ginneries.
Coffee	Coffee Board	Organizes auctions for sale to licensed buyers.

	KNCU	Coffee co-operative of the Chagga.
	Tanganyika Co-operative Trading Agency (TACTA)	Co-operative sale of coffee produced on small farms outside of the KNCU. Colvale & Co. Managing Agent.
	Tanganyika Coffee Growers' Association	Organization of the large coffee producers. Processes coffee together with the KNCU, sales organization.
Cashew nuts	Southern Region Cashew Board	Organizes sales at auctions, grades. Handles 80% of the supply.
	Co-operatives	Purchase the complete supply from the producers, deliver it to the Board. Handle 10—15% of the production.
Tea	Tea Board	Functions as adviser to the government.
	Tanganyika Tea Growers' Association	Producers' organization for consultation on matters of common interest and for the promotion of research.
Pyrethrum	Pyrethrum Board	Organizes sales from the producer to the processor, advises the government, finances research.
Seed beans	Seed Board	Advises the government, arranges contracts between foreign buyers and the producers, promotes research.

Source: Taken from: Tanganyika Commodity Marketing and Price Stabilization. Appendix A. Dar-es-Salaam, 1962, Mimeo.

2. Market Controls

a) Wheat and Sugar

The Ministry of Commerce and Industry regulates the markets for sugar and wheat. As a British dependent territory, Tanganyika was made partner to the International Wheat Agreement, the International Sugar Agreement and the Commonwealth Sugar Agreement. In addition, there are agreements with Kenya, regulating the import of wheat from Kenya, and the prices and distribution of milling quotas for Tanganyika's mills.

The government is the sole importer of sugar, wheat and wheat flour. Sugar prices to the producers are taken from the mean between the prices under the Commonwealth Sugar Agreement and those of world market. The consumer price corresponds to a mean between the home produced and imported sugar prices. Price control is limited to the wholesale trade. Wholesale prices for wheat are also subject to control. The aim is to establish the same price for producers in Tanganyika and in Kenya.

The Ministry of Commerce and Industry influences the prices for maize and rice by regulating the quantities imported. Duties are imposed. Up to 1961, maize carried an import duty of 3/5 d. shs/cwt. The bad harvest of

1961 led to the import duty being lifted. New import restrictions came into force in the middle of 1962. The rice producers are protected by an import duty of 10 shs/cwt.

b) Cattle Markets

With few exceptions cattle markets in Tanganyika are run by the Local Authorities. The original organization of these markets was carried out by the Veterinary Department, which is still responsible for the control of the stock-routes and quarantines. Cattle, sheep and goats are sold by auction on local cattle markets. The largest buyer, buying up 25—30 per cent of the supply, is Tanganyika Packers Ltd., a company of which more than half is state-owned and which is managed by the private shareholder, the Liebig Company. A predominantly state-owned enterprise is, therefore, the dominant factor in the market. The price policy of this company is dictated by the world market for canned meat. In consequence, Tanganyika Packers are not able to pay the same prices as local butchers. They buy up the stock left by other buyers.

c) Milk and Dairy Products

The organization of dairy markets is still in its infancy. It is planned to set up Area Dairy Boards. They are to control the whole dairy industry, from the producer to processor and the wholesale trade. Considerable difficulties can be expected. The milk does not meet the requirements of hygiene. The supply is irregular and concentrated on a few areas with seasonal surpluses. In addition, there is keen price competition from milk powder, tinned milk etc. from Holland and Denmark.

The first project is to be started in the Moshi/Arusha district, with the help of UNICEF. It is planned to establish a dairy and to make the sale of milk through it compulsory. Seasonal surpluses are to be processed into dairy products. If milk production rises rapidly the manufacture of milk powder will be considered.

Bibliography

1. The bibliography is confined to studies relating specifically to Tanganyika, in whole or in part.

2. The material listed is that which appears useful for an assessment of the conditions pertaining to agriculture and its development. Studies of a purely technical character are not included.

3. The bibliography commences with the 1920's. The many, in part still topical studies from the time prior to 1914 are not listed. We would refer to the bibliography of the former library of AMANI, now available at Muguga.

4. Numerous little known but particularly informative manuscripts are kept by the Ministries in Dar-es-Salaam and in the offices of the Regional and District Agricultural Officers. We would refer in particular to the "Agricultural District Books".

5. Almost all the publications and mimeographs listed can be found in the libraries of the following:

a) Agricultural Department in Dar-es-Salaam.
b) Research stations at Ukiriguru, Tengeru and Kilosa (Ilonga).
c) The "White Fathers" at Nyegezi near Mwanza.
d) The East Africa Agricultural Research Institute at Muguga, near Nairobi.

6. Abbreviations.

n. d.	No date.
MS.	Manuscript.
Mimeo	Mimeography.
OFC	Overseas Food Corporation.
H.M.S.O.	Her (His) Majesty's Stationery Office.
J. of Afr. Adm.	Journal of African Administration.
T.N.R.	Tanganyika Notes and Records.
E.A.A.J.	East African Agricultural Journal.
C.D.C.	Colonial Development Corporation.
Emp. Cot. Grow. Rev.	Empire Cotton Growing Review.
FAO	Food and Agricultural Organization.
Trop. Agr.	Journal of Tropical Agriculture.
Emp. J. of Ex. Agr.	Empire Journal of Experimental Agriculture.
T.C.N.	Tanganyika Coffee News.
Col. Rev.	Colonial Review.
Afr. Aff.	African Affairs.
E. A. Econ. Rev.	East African Economic Review.
E. Afr. Med. J.	East African Medical Journal.
C.C.T.A.	Commission for Technical Cooperation in Africa South of the Sahara.

ACHTNICH, W., B. GRZIMEK, G. KRAGH, A. VON NAGY, R. SCHMIDT-LORENZ, and O. SPLETT: Landschaftsplanung in Afrika. Sept. 1961, Mimeo.

ADAM, V.: Social composition of Isanzu villages. In: E. A. Institute of Social Research, Kampala, Conference, Jan. 1962, Mimeo.

AKEHURST, B. C.: Flue-cured tobacco in Tanganyika. Dep. of Agr. Bull. No. 4.

ANDERSON, B.: A survey of soils in the Kongwa and Nachingwea districts of Tanganyika. Univ. of Reading 1957.

ANKERMANN, B.: Das Eingeborenenrecht, Sitten und Gewohnheitsrechte der Eingeborenen der ehemaligen deutschen Kolonien in Afrika und der Südsee. Stuttgart 1929.

APIYO, T.: The agricultural aspirations of the Ifakara people. Dep. of Agr. 1961 (?). Mimeo.

ARMITAGE-SMITH, S.: Financial mission to Tanganyika. H.M.S.O. 1932.

AUCKLAND, A. K.: Soya beans in a mechanized farming system in Tanganyika. In: Trop. Agr. Vol. 37, No. 3, July 1960.

— and F. M. SPENCER: Soyabeans, Dep. of Agr. Bull. No. 14, Dar-es-Salaam 1961.

BAKER, E. C.: Report on social economic conditions in the Tanga Province. Dar-es-Salaam 1934.

— Report on the administration and social conditions in Uzinza. Mwanza 1935 (?) MS.

Baker, R. E. D.: Report on a visit to British East Africa to study the cultivated and wild bananas. I.C.T.A. Trinidad 1948.

Barclays Bank: Tanganyika. An economic survey. London 1955.

Barnes, A. C.: The present position and development of the sugar-cane industry in East Africa. Nairobi 1952, Mimeo.

Beck, R. S.: An economic study of coffee-banana farms in the Machame Central Area, Kilimanjaro District, Tanganyika, 1961, Dar-es-Salaam 1963, Mimeo.

Beidelman, Th. O.: Beer drinking and cattle theft in Ukaguru: Intertribal relations in a Tanganyika Chiefdom. In: American Anthropologist, Vol. 63, June 1961.

Beidelman, F. O.: Umwano and Ukaguru students' association. In: Anthropos Vol. 56, No. 5 and 6, 1961.

Bell, M. F.: Tea cultivation. Dep. of Agr. Pamphlet No. 3, Dar-es-Salaam 1930 (?).

— Tea planting prospects in the South-Western Highlands of Tanganyika. London, 1928.

Bellamy, J.: Tanganyika cotton industry. In: Emp. Cot. Grow. Rev. Vol. XXXVII, April 1960.

Bennet, A. L. B.: A short account of the Kilimanjaro Native Co-operative Union Ltd. In: E.A.A.J. Vol. XII, July 1946.

Berger, H.: Tanganyika, Bevölkerungs- und wirtschaftsgeographische Übersicht eines Entwicklungslandes. in: Mitt. d. Östr. Geograph. Gesellschaft 104. 1962: 1/2.

Blohm, W.: Die Nyamwezi. 1. Land und Wirtschaft. Hamburg. 1931. 2. Gesellschaft und Weltbild. Hamburg, 1933.

Bösch, F.: Les Banyamwezi. Münster, 1930.

Bomani, P.: Baumwollgenossenschaften in der Seeprovinz Tanganyikas. in: Internat. Genoss. Rundschau Bd. 49, Nr. 2, Febr. 1956.

Braun, K.: Überblick über die im Laufe der Jahre in Amani angebauten und beobachteten Kulturpflanzen besonders der Eingeborenen. in: Kolon. Rundsch. 26. Jg. Heft 6, 1935.

Brewin, D. R.: Teaching the Tanganyika agriculturist. In: Corona Vol. 13, 1961.

Bridger, G. A.: Planning land settlement schemes. In: ECA—FAO Agricultural Economics Bulletin for Africa, No. 1, 1962, Mimeo.

— Peasant tea production in Tanganyika. ECA—FAO, April 1961, Mimeo.

Brooke, H. C.: A Geographical Study of Famines and Acute Shortages of Food in Tanganyika. Portland State College, 1963.

Brown, G. G., and A. B. Hutt: Anthropology in action. An experiment in Iringa District. London 1935, p. 134 ff.

Brown, K. J.: Effect of early rainfall on the response of cotton to nitrogen. In: Emp. Cot. Grow. Rev. Vol. XXXIX, No. 3, July 1962.

— Rainfall, tie-ridging and crop yields in Sukumaland. in: Emp. Cot. Grow. Rev., Vol. XXXI, Jan. 1963.

Bryceson, D.: Tanganyika, the challenge of Tomorrow. In: Africa in Transition. Princeton 1961.

Bukoba Native Coffee Board: Annual Report.

Bunting, A. H.: Agricultural research in the groundnut scheme. 1947—1951. In: Nature, Vol. 168, 1951.

— Some biological problems arising from large-scale mechanised agriculture in East Africa. In: Comm. B(h)20. Afr. Reg. Sci. Johannesburg, 1949.

— Technical aspects of the East African groundnut organization. In: Ann. App. Bio. 37, 4, 1950.

— Land development and large scale food production in East Africa by the Overseas Food Corporation. In: Econ. Bot. 6. 1. 1952.

Byamungu, A.: An appraisal of the general agriculture of the Nguu. Handeni 1960, MS.

Calton, W. E.: Experimental pedological map of Tanganyika. In: C.C.T.A. Conference Leopoldville, August 1954.

Cameron, J. F., B. W. Keulen, R. van Toxopeus, and A. H. Bunting: Pilot scale crop costings on groundnuts, sorghum and maize at Urambo experimental station. Tang. Terr. 1949—50. In: Emp. J. of Exp. Agr. 19, 1951.

Catalogue of the Amani Library: December 1934, Amani.

Césard, E.: Les Muhaya. In: Anthropos Vol. XXX, XXXI and XXXII, 1935—37.

Chablani, M.: Outline plan for the development of the Ruvu Basin. FAO, Dar-es-Salaam 1961 (?), Mimeo.

Chalmers-Wright, F.: Distribution and consumption of commodities among Africans in Nyasaland and Tanganyika, H.M.S.O. London, 1955.

Charpentier, A. E.: Second report on the development of the dairy industry with particular reference to the Northern Province. FAO, Rome 1960.

Charron, K. C.: The welfare of African labour in Tanganyika. Dar-es-Salaam 1944.

Charter, C. L.: Report on environment conditions prevailing in Block "A", Southern Province, Tanganyika, Accra, 1958.

Chesam, Lord: Settlement in Tanganyika. In: Journal of the Royal Africa Society. 37., April 1937.

Chidzero, B. T. G.: Tanganyika and international trusteeship. Oxford Un. Press 1961.

Childs, A. H. B.: Cassava in Tanga Province. In: E.A.A.J. Vol. XXIII, Jan. 1958.

— Copra in the Tanga Province of Tanganyika. In: E.A.A.J. Vol. XXIII, April 1958.

— Cassava. Min. of Agr. Bulletin No. 15, Dar-es-Salaam 1961.

— and C. G. Groom: Coconut — cattle an exercise in agricultural extension. Tanga, 1962 MS.

Clegg, C. G.: Notes on chickpea and its cultivation in the Lake Province. In: E.A.A.J. Vol. XIII, July 1947.

— Native foodstuffs in Tanganyika: The preparation and use of local foodstuffs in the Shinyanga District of Sukumaland. In: Trop. Agr. Vol. XXII, Nr. 2, 1945

— Ukerewe Island. The agricultural position, its potentialities and difficulties. 1947 MS.

Cocking, W. P., and R. F. Lord: The Tanganyika Agricultural Corporation's farming settlement scheme. In: Trop. Agr. Vol. 35, April 1958.

Coffee Board of Tanganyika: Coffee of Tanganyika. London, 1959 (?).

Coffee in Bukoba: Bukoba, 1950.

Collinson, M. P.: Report on Bukumbi survey. Nov. 1961 to Nov. 1962. Ukiriguru. Mimeo.

— Farm management survey, No. 2, Western Research Centre. Usamao, Sukumaland, 1963. Mimeo.

C.C.T.A. Proceedings: Symposium on the conservation of nature and natural resources in modern African States. FAO—UNESCO. Arusha, 1961.

— Conference on the mechanisation of agriculture. Uganda 1955. Tanganyika: Fuggles-Couchman, N. R., Hubbard, H. E., Constantinesco, I., Paxton, R. H., Schultz, H. E., Minto, D., Foster, J., Haswell, M. R., Forbes, A. P. S., Mimeo.

Conference on the medical, agricultural and veterinary aspects of food production in East Africa. In: E.A.A.J., Vol. XX, July 1954.
CONSTANTINESCO, I.: Research and development towards improved farm implements for peasant farmers in the tropics. In: Journal of the Institute of Agr. Engineers (in print).
COOPER, H. M.: Farm and estate planning in Tanganyika. Min. of Agr. Bull., No. 12, Das-es-Salaam, 1958 (?).
Cooper & Brothers: Report on cotton industry in Tanganyika. Min. of Nat. Res. Dar-es-Salaam, Febr. 1959.
Co-operative Movement in Tanganyika. Dar-es-Salaam 1961 (?), Foreword P. BOMANI.
CORY, H.: The indigenous tribal structure of Sukumaland and its relation to constitutional reforms. o. J. MS.
— History of Bukoba District. In: T.N.R., No. 54, 1960.
— The indigenous political system of the Sukuma. E. A. Studies No. 2, Dar-es-Salaam, 1954.
— Land tenure in Bukaria. In: T.N.R. No. 23, 1947.
— The necessity for co-ordination of economic and political development. In: T.N.R., No. 30, Jan. 1951.
— Sukuma law and custom. Oxford, 1953.
— and M. M. HARTNOLL: Customary law of the Haya Tribe, Tanganyika Territory. London, 1945.
CROSSE-UPCOTT, A. R.: Labour migration in Liwale (Southern Province) n. d. Mimeo.
— Ngindo famine subsistence. In: T.N.R., No. 50, 1958.
CULWICK, A. T.: The hoe in Ulanga. In: Man, No. 39, 1934.
— The labourer and his hire. In: T.N.R. No. 17, 1944.
— Land tenure in Tanganyika Territory. Dep. of Agr. 1939 (?) Mimeo.
— and G. M. CULWICK: Ubena of the rivers. London, 1935.
CULWICK, G. M., and A. T. CULWICK: A study of factors governing the food supply in Ulanga, Tanganyika Territory. In: E. Afr. Med. J. May 1939.
DAMES, T. W. G.: The soil of the Pangani Valley. FAO. Report No. 970, 1959.
DAVIES, D. A., D. HEPBURN, and H. W. SANSOM: Report on experiments at Kongwa on artificial stimulation of rain. In: Memoirs of E. Afr. Meteor. Dep. No. 2, 1951.
— Experiments on artificial stimulation of rain in East Africa. In: Nature, No. 174, 1954.
DEBENHAM, F.: Report an the water resources of the Bechuanaland Protectorate, Northern Rhodesia, the Nyasaland Protectorate, Tanganyika Territory, Kenya and Uganda Protectorates. H.M.S.O. London, 1948.
Der landwirtschaftliche Dienst und das landwirtschaftliche Versuchswesen in den deutschen Schutzgebieten. In: Tropenpflanzer, Juli/August 1922.
DIGBY, M.: Success for the Chagga. Fabian Col. Bureau. May 1951.
DOBSON, E. B.: Comparative land tenure of ten Tanganyika tribes. In: J. of Afr. Adm., April 1954.
— Land tenure of the Wasambaa. In: T.N.R. No. 10, 1940.
DOGGET, H. S.: Sorghum improvement in Tanganyika. In: E.A.A.J. Vol. XXV, July 1959.
DOLFI, D.: Notes on the Kalemawe scheme. Dar-es-Salaam 1957. MS.
DOOP, J. E. A.: Prospects for the utilization of sisal waste. In: E.A.A.J. Vol. XV, July 1949.

DOW-SMITH, G. T.: British East Africa, an overseas economic survey. London, 1953.

DRENNAN, D. H.: Soil conservation and land use in steep highland areas. In: Conf. of the E. A. Reg. Com. for Conservation and Utilisation of the Soil. July 1956. Mimeo.

DUDBRIDGE, B. J.: Some notes on the duck and geese of Usukuma. In: T.N.R. No. 31, 1951.

DUDGEON, J. B., J. B. MAYNE, and C. A. WALDRON: Prospects of establishing a rubber growing industry in the Ulanga District of Tanganyika. C.D.C. Dar-es-Salaam, 1961.

DUFF, P. C.: Note on "Land and politics among the Luguru of Tanganyika". In: T.N.R. No. 56, 1961.

DUNDAS, C.: Kilimanjaro and its people. London, 1924.

DUNN, R. P.: Cotton in British East Africa. National Cotton Council. Memphis. Tennessee 1949.

DUTHIE, D. W.: Mound cultivation (Matengo). In: E.A.A.J. Vol. XXVI, Oct. 1959.

DYER, H. W.: Report on the situation of the co-operative movement in Tanganyika. 1960 MS.

East African Agricultural Research Station, Amani: Annual reports to 1947.

East African Common Services Organization: Quarterly economic and statistical bulletin. Nairobi.

— Research programmes. Nairobi (annually).

East African High Commission: African population of Tanganyika Territory. E. A. Stat. Off. Nairobi, 1951.

— Annual report. Nairobi, to 1961.

— Agricultural census 1958. Stat. Dep. Tang. Unit. 1959.

— Annual trade report. (Since 1962: E. A. Common Services Org.)

— Census of large scale commercial farming in Tanganyika. Stat. Dep. Tang. Unit., June 1961.

— Agriculture and Forestry Research Organization. Record of research. Nairobi (annually).

— East African Veterinary Research Organization. Annual report. Nairobi.

— The external trade of East Africa, indices 1954—1958. Nairobi, 1960.

— Household budget survey of Africans living in Dar-es-Salaam 1956—1957. Stat. Dep. Tang. Unit.

— Hydrology and water resources of British Eastern and Central Africa. Nairobi, 1951. Mimeo.

— The pattern of income, expenditure and consumption of African workers. a) Dar-es-Salaam, August 1950 b) Tanga, February 1958 c) Mwanza, September 1958. Stat. Dep. Tang. Unit. Mimeo.

— Quarterly economic and statistical bulletin. Nairobi, to 1961.

— Report on the analysis of the sample census of African agriculture. 1950 (revised.) 1953.

— Report of the commission on the most suitable structure for management direction and financing of research on the East African basis. Nairobi 1961.

East African Railways and Harbours: The economy of East Africa. A study in trends. London 1955.

EDEN, T.: The tea industry of East Africa. In: World crops, Vol. 6, No. 5, May 1954.

— Report on the possible extension of tea cultivation in Tanganyika. Dar-es-Salaam, 1956.

Eliasberg, G.: Report on settlement schemes. Dar-es-Salaam 1962, Mimeo.

Edwards, D. C., W. J. Eggeling, and P. J. Greenway: East African vegetation map. East African Agriculture and Forestry Research Organization 1948.

Empire Cotton Growing Corporation: 1921 to 1950. In: Emp. Cot. Grow. Rev. Vol. XXVIII, No. 1, 1951.

— Progress and reports from experiment stations, Tanganyika Territory, Lake Province (annually) (see also: "Coastal Provinces").

Evans, A. C.: A study of crop production in relation to rainfall reliability. In: E.A.A.J. Vol. XX, April 1955.

— and L. R. Wilson: Responses to fertilizer under farming conditions in the Southern Province of Tanganyika. In: E.A.A.J. Vol. XXII, July 1956.

FAO: Milk development possibilities in East Africa. Report on an exploratory visit. Rome, 1958.

— Report to the government of Tanganyika in a dairy survey of the Arusha/Moshi Area. 1959.

— The Rufiji Basin, Tanganyika. Rep. No. 1269. Vol. I: General report; Vol. II: Hydrology and water resources; Vol. III: Water control; Vol. IV: Irrigation development; Vol. V: Mbarali irrigation scheme; Vol. VI: Geology (in preparation); Vol. VII: Soils of the main irrigable areas. Dar-es-Salaam (?), 1961 (?).

— An outline plan for the development of the Ruvu Basin. 1961, Mimeo.

— *and C.C.T.A.:* History and present status of extension work in Tanganyika. In: Agricultural Extension Development Centre. February 1962, Arusha. Mimeo.

FAO/UNICEF: Draft report on the nutrition seminar held at Arusha. August 1962.

Fabian Colonial Bureau: Cooperation in the colonies. London 1945 (see page 20 ff.).

Ford, J.: Rinderpest. In: T.N.R. No. 36, 1954.

Ford, V. C. R.: The trade of Lake Victoria. East African Studies No. 3, Kampala 1955.

Fosbrooke, H. A.: An administrative survey of the Masai social system. In: T.N.R. Vol. 26, 1948.

— Government sociologist in Tanganyika. In: J. Afr. Adm., Vol. IV, July 1952.

— A sociological survey of the Masai of Tanganyika Territory. n. d. MS.

Frankel, S. H.: The economic impact on under-developed societies. Oxford, 1953.

Freimann, M., and N. Simansky: Report and estimates on proposed Mbarali River 5000 acres — pilot irrigation scheme. Dar-es-Salaam, 1957, Mimeo.

French, M. H.: Development of livestock in Tanganyika Territory. In: Emp. Cot. Grow. Rev., Vol. X, No. 39, July 1952 and Vol. XII, No. 48, Oct. 1944 and in: Emp. J. of Exp. Agr. Vol. IV, No. 13, Jan. 1936, Vol. VI, No. 22, April 1938, Vol. VI, No. 23, July 1938, Vol. VIII, No. 29, Jan. 1940.

— and H. J. Lowe: Clarified butter production. Vet. Dep. Tang. Dar-es-Salaam 1936.

— Feeding of dairy cows. Vet. Dep. Dar-es-Salaam 1939.

— The liveweight development of certain shorthorned zebu cattle in Tanganyika Territory. In: Trop. Agr. Vol. 16, 1951.

— Factors affecting animal nutrition in Tanganyika. In: E.A.A.J. Vol. XVI, April 1951.

— Notes on the hides and skins industry. In: E.A.A.J., Vol. XI, No. 4 and XII, No. 1.

Friedland, W. H.: A Study of Trade Union Development in Tanganyika. University of California, Berkeley 1963.

Fuggles-Couchman, N. R.: Agricultural Change in Tanganyika, 1945—1960. In: Food Research Institute Studies in Tropical Development (in print).
— Large scale wheat production at Oldeani, Tang. Terr. In: E.A.A.J., Vol. IX, April 1944.
— Wheat production in Tanganyika. Min. of Agr. Bull. No. 9. Dar-es-Salaam, n. d.
— and G. B. Wallace: Cultivation and diseases of wheat. Tang. Terr. Dep of Agr. Dar-es-Salaam, 1945.
— Agricultural problems in Tanganyika. In: Corona, Vol. 12, No. 12, December 1960.
Gaitskell, A.: Report on land tenure and land use problems in the Trust Territories of Tanganyika and Ruanda-Urundi. FAO 5/2/951, Febr. 1951, Mimeo.
Gale, G. W.: Food balance sheets for the African population of East Africa. In: East Afr. Med. J. 37, 410 (1960).
Gethin-Jones, G. H.: Some notes on a reconnaissance soil survey of Sukumaland. 1953, MS.
Gibbs, A. L., and partners: Report on Smith Sound scheme. London, 1955/56.
— and partners: Water resources survey of the Nile Basin in Tanganyika. London, 196.
Gilbert, S. M.: The coffee research experimental station: A brief survey of the first ten years' work. In: Emp. J. Exp. Agr., July 1945.
— The coffee research and experimental station, Tanganyika Territory. In: Emp. J. of Exp. Agr., Vol. XIII, No. 51, July 1945.
— The multiple stem system of growing coffee arabica. Dep. of Agr. Pamphlet No. 39, Dar-es-Salaam, 1945.
— The objects and scope of the coffee research and experimental station. Dep. of Agr. Pamphlet No. 15. Dar-es-Salaam, 1935.
Gillman, C.: South-West Tanganyika Territory. In: Geogr. J., Vol. LXIX, No. 2, Febr. 1927.
— A geographer's retrospect and note on the Mpwapwa land-usage experiments and achievements. In: E.A.A.J., Vol. VII, Jan. 1943.
— Problems of land utilization in Tanganyika Territory. In: South Afr. Geogr. J., Vol. XX, April 1938.
— Man, land and water in East Africa. In: E.A.A.J., Vol. III, March 1938.
— Population problems of Tanganyika Territory. In: E.A.A.J., Vol. XI, Oct. 1945.
— Reconnaissance survey of the hydrology of Tanganyika in its geographical settings. Water Consultant's Rep. No. 6, Dar-es-Salaam, 1940.
— The geography and hydrography of the Tanganyika Territory part of the Ruvuma Basin, 1943.
— Vegetation-types map of Tanganyika Territory, 1949. Amer. Geogr. Soc., New York, 1949.
— White settlement in East Africa with special regard to Tanganyika Territory. In: Geographical Rev., Vol. 32, Oct. 1942.
— Problems of land utilization in Tanganyika Territory. In: S. Afr. Geogr. J., Vol. XX, April 1938.
— Eine interessante Form der Landzerstörung durch den Menschen. In: Wiss. Ver. des Deutsch. Mus. f. Länderkunde, Leipzig 1939.
— A hydrographic reconnaissance into parts of Masailand. Water Consultant's Rep. No. 1, 1938.
— A bibliography of Kilimanjaro. In: T.N.R., No. 18, 1944.
— Bush fallowing in the Makonde Plateau. In: T.N.R., No. 19, 1945.
Glendon-Hill, A.: Oil plants in East Africa. In: E.A.A.J., Vol. XII, No. 3, Jan. 1947.

GLOVER, J.: Environment and the growth of potatoes in tropical East Africa. In: Emp. J. of Exp. Agr. Vol. XV, No. 57, Jan. 1947.
GLYNN, F. J.: Africanization and job analysis in Tanganyika. In: J. of Local Adm. Overseas. Vol. II, No. 3, July 1963.
GRANT, C.: The Uha in Tanganyika Territory. In: Geogr. J., Vol. LXVI, No. 5, 1925.
GRAY, F. R.: Economic exchange in a Sonjo Village. In: Markets in Africa. Northwestern Univ. Press. 62.
— Sonjo bride-price and the question of African "wife purchase". In: American Antropologist, Vol. 62, Febr. 1960.
GREENWAY, P. J.: Origin of some East African food plants. In: E.A.A.J., Vol. X, 1944/45 (s. a. STUHLMANN, F.: Beiträge zur Kulturgeschichte Ostafrikas, Berlin 1909).
— G. B. WALLACE, M. M. WALLACE, and E. V. R. KHOMO: The papaw. Dep. of Agr. Pamphlet No. 52, Dar-es-Salaam, 1953.
GRIFFITHS, J. H.: Climatic zones of East Africa. In: E.A.A.J., Vol. XXIII, Jan. 1957.
GRIFFITHS, J. E. S.: Masai cattle auction. In: T.N.R. No. 6, 1938.
— Notes on land tenure and land rights among the Sonjo of Tanganyika Territory. In: T.N.R. No. 9, 1940.
GUILLEBAUD, C. W.: An economic survey of the sisal industry of Tanganyika. London, 1958.
GUISE-WILLIAMS, C.: Memorandum on village organization in Usukuma. n. d. MS.
GULLIVER, P. H.: Alien Africans in Tanga region. Dar-es-Salaam 1956.
— Incentives in labour migration. In: Human Organization, 19. 3. 1960.
— Labour migration in a rural economy. A study of the Ngoni and Ndendeule of Southern Tanganyika. E. A. Studies No. 6, Kampala, 1956 (?).
— Land tenure and social change among the Nyakyusa. E. Afr. Studies No. 11, Kampala, 1958.
— Nyakyusa labour migration. In: Rhodes-Livingstone Journal, No. XXI, 1957.
— The evolution of Arusha trade. In: Markets in Africa. Northwestern Univ. Press. 1962.
— The population of the Arusha Chiefdom: High density area in East Africa. In: Rhodes-Livingstone Journal, Vol. XXVIII, Dec. 1960.
GUNN, D. L.: Locust control by aircraft in Tanganyika. Pretoria, 1948.
GUNN, J. S.: The responses of wheat, tobacco und pyrethrum to fertilizers in the Southern Highlands of Tanganyika. In: Proceedings, Muguga, 1953, Mimeo.
GUTMANN, B.: Recht der Dschagga. Arbeiten zur Entwicklungspsychologie. No. 7, Berlin, 1926.
HAARER, A. E.: Production of arabica coffee. Dar-es-Salaam, 1929.
HAILY, L.: Native administration in the British African Territories. London, 1956.
HALCROW, W., and partners: Report on reconnaissance of the Pangani River Basin. London, 1950.
HALDEMANN, E. G.: Recent landslide phenomena in the Rungwe volcanic area, Tanganyika. In: T.N.R. No. 45, Dec. 1956.
— (Translation): Description of Lake Victoria and the Wembere savanna irrigation scheme. In: T.N.R. No. 47—48, 1957 (s. LUDIN A. and THOMA, E.: Die Wasserwirtschaft in Afrika).
HALL, R. DE Z.: Irrigation in Bugufi, Tanganyika Territory. In: Man. 20, 1939.
— Local migration in Tanganyika. In: African Studies, Vol. IV, No. 2, June 1945.
— and H. CORY: Study of land tenure in Bugufi 1925—1944. In: T.N.R., No. 24, 1947.

HANCOCK, D. H.: Cultivation of hibiscus can. in the Nachingwea area of Tanganyika Agricultural Corporation. 1958, Mimeo.

HARRIS, J. S.: Report on Uluguru land usage scheme. n. d. MS.

HARRIS, W. V.: Native bee-keeping in Tanganyika Territory. In: Trop. Agr., Vol. IV, No. 8, Aug. 1932.

— Native methods of food storage in Tanganyika. In: E.A.A.J., Vol. VI, Jan. 1941.

HARRISON, E.: Soil erosion memorandum. Tang. Terr. Milbank 1937.

HARTLEY, B. J.: Land tenure in Usukuma. In: T.N.R. No. 5, 1938.

HARTLEY, B. T. H.: Land tenure and settlement in Usukuma, with special reference to the Kwimba and Maswa Districts. Mwanza, 1934, MS.

HARTNOLL, A. V., and N. R. FUGGLES-COUCHMAN: The mashokora cultivation of the coast. T.N.R. No. 3, 1937.

HASSEL, K.-U. VON: Entwicklungsländer Ostafrikas. Vortrag, Bad Godesberg 1960. MS.

HATCHELL, G. W.: The Angoni of Tanganyika Territory. In: Man. Vol. XXXV, May 1935.

— Resettlement in the areas reclaimed from tsetse-fly. In: T.N.R., No. 53, 1959.

HEADY, A. F.: Range management in East Africa. Nairobi, 1960.

HEDBERG, O.: Vegetation belts of the East African mountains. Svensk Botanisk Tidschrift. Vol. 45, No. 1, 1951.

HEMY, C. D.: Onions production. Dep. of Agr. Bull. No. 2, Dar-es-Salaam, 1960.

H.M.S.O.: Report by agricultural advisers to the Secretary of State for the Colonies on his recent visit to Tanganyika Territory. 1937.

— Papers relating to the question of the closer Union of Kenya, Uganda and the Tanganyika Territory, London 1931.

— Amani, East Africa Agricultural Research Institute. Reports (ending 1945).

— A plan for the mechanized production of groundnuts in East and Central Africa. Cmd 7030, London, 1947.

— East African Royal Commission 1953—55. Report. London, 1955.

— Colonial Development Corporation. Report and Accounts (annually).

— Dispatches from the governors of Kenya, Uganda and Tanganyika and from the administrator, East African High Commission, commenting on the East African Royal Commission, 1953—55. Report. London, 1956.

— The future of the Overseas Food Corporation. London, 1954.

— Land and population in East Africa. London, 1952.

— Notes on some agricultural development schemes in Africa and Aden. 1st. Revise 1953. 2nd Revise 1953. London. Mimeo.

— Report of survey of problems in the mechanization of native agriculture in tropical territories. London, 1950.

— Report of the Sorghum Mission to certain British African Territories. London, 1951.

— Tanganyika. Annual Report (ending 1961). London.

HICKMAN, G. M., and W. G. H. DICKENS: The land and peoples of East Africa. London, 1960.

HILL, J. F. R., and J. P. MOFFET: Tanganyika: a review of its resources and their development. London, 1955.

HINDORF, R. J.: Der Sisalbau in Deutsch-Ostafrika. Berlin, 1925.

HITCHCOCK, E.: The sisal industry of East Africa. In: T.N.R., No. 52, 1959.

HOLLOWAY, J. W. T.: De-stocking and culling in Tanganyika. In: E.A.A.J., Vol. XX, Jan. 1954.

HORNE, H.: The extension of cotton cultivation in Tanganyika Territory. Emp. Cot. Grow. Corp. London 1922.

HORNBY, H. E., and H. J. VAN RENDSBURG: The place of goats in Tanganyika's farming system in deciduous bushland formation. In: E.A.A.J., Vol. XIV, Oct. 1948.

— Overstocking in Tanganyika. In: E.A.A.J., Vol. I, March 1936.

HUPPERTZ, J.: Die Eigentumsrechte bei den Masai. In: Anthropos, Vol. 54, 1959.

International Bank for Reconstruction and Development: The economic development of Tanganyika. Dar-es-Salaam, 1960.

JACK, D.: The agriculture of Rufiji District. 1957, Mimeo.

JACK, D. T.: Report on the methods of determining wages. Dar-es-Salaam, 1959.

JAEGER, F.: Zur Geographie der ländlichen Siedlung in Ost-Afrika. In: Die ländlichen Siedlungen verschiedener Klimazonen. Breslau, 1933.

JERRARD, R. C.: The tribes of Tanganyika, their districts, usual dietary and pursuits. Dar-es-Salaam, 1936.

JERVIS, T. S.: A history of robusta coffee in Bukoba. In: T.N.R. No. 8, 1939.

JOHNSTON, B. F.: The choice of measures for increasing agricultural productivity. A survey of possibilities in East Africa. In: Tropical Africa (in print).

JOHNSTON, P. H.: Land tenure on Kilimanjaro. In: T.N.R. No. 21, 1946.

JONES, W. O.: Manioc in Africa. Stanford 1959 (Tanganyika p. 230).

JOSHI, N. R., E. A. MCLAUGHIN, and R. W. PHILIPPS: Tanganyika shorthorn zebu. In: Types and Breeds of African Cattle, FAO 1957.

KANTHACK, F. E.: Report on the control of the natural water of Tanganyika and framework of a water law on which such control should be based. Dar-es-Salaam, 1938.

KAPLAN, B.: New settlement and agricultural development in Tanganyika. Beèr Sheva, Israel, 1961. Mimeo.

KAUFMANN, R., and R. PILSON: Notes on hand-clearing as carried out by the O.F.C. in the Southern Province between May—November 1951. Mimeo.

KAYSER, K.: Die Eingeborenenarbeit als Problem der Bevölkerungs- und Wirtschaftsstruktur Deutsch-Ostafrikas. In: Geographische Zeitschrift 45 (4), 1939.

KELLER, W.: Studie zur Ernährung bei zwei Stämmen in Nord-Tanganyika. Max-Planck-Institut für Ernährungsphysiologie. Dortmund 1962. Mimeo.

KENT, A. W.: Report on the services to be administered by local authorities in Tanganyika and consequential financial arrangements. Dar-es-Salaam, 1962.

Kilimanjaro Native Co-operative Union Limited (KNCU): Annual Reports.

KING, J. G. M.: Memorandum on soil erosion and reafforestation in Shinyanga District. 1937, MS.

— Memorandum on locust destruction. Dep. of Agr., Dar-es-Salaam, 1934.

Kingolwira Peasant Settlement Scheme. In: Emp. Cot. Grow. Rev. No. 15, 1938.

KIRBY, A. J.: The progress of cotton growing in Tanganyika Territory. In: Emp. Cot. Grow. Rev., Vol. VII, No. 3, July 1930.

KIRPATRICK, F. W.: The climate and ecoclimate of coffee plantation in East Africa. Amani Memoirs, 1935.

— The ecology of coffee plantations. In: E.A.A.J., Vol. I, 1936.

KUENZLER, A.: Ein Umsiedlungsprojekt für den Stamm der Waarusha. Arusha, 1961. Mimeo.

LACEY, G., and R. WATSON: Report on rice production in east and central African territories. 1948, Col. Res. No. 246, London, 1949.

LANG, G. O., and M. B. LANG: Problems of social and economic change in Sukumaland, Tanganyika. In: Anthropological Quarterly, Vol. 35, No. 2, April 1962.

LANGE, H.: Kultur und Aufbereitung von röhrengetrocknetem Virginia-Tabak im Iringa-Bezirk, Deutsch-Ostafrika. In: Tropenpflanzer, Juli 1940.

Leaky, E. A., and N. V. Rounce: The human geography of the Kasulu District, Tanganyika. In: Geography, 1933, p. 292.
Le Mare, P. H.: Crop responses to fertilizers and manures in East Africa. In: E.A.A.J., Vol. XIX, July 1953.
Leubuscher, Ch.: Marketing schemes for native grown produce in African territories. In: Journal of the Intern. Inst. of Afr. Lang. and Cult., Vol. 12, No. 2, 1939.
— Tanganyika Territory. A study of economic policy under the mandate. London 1944.
Leutenegger, F.: Provisional reports of soils of sisal estates in Tanganyika. Sisal Res. Sta. Bull. No. 12—19, 1955—56.
Lichtenheld, G.: Über die Bekämpfung der Tierseuchen im ehemaligen Deutsch-Ostafrika. In: Tropenpflanzer, Jan./Febr. 1923.
Liebenow, J. G.: Chieftainship and local government in Tanganyika. A study of institutional adaptation Diss. Evanston, Illinois.
— The establishment of legitimacy in a dependency situation. A case study of the Nyaturu of Tanganyika. In: African studies. Vol. 20, No. 1, 1961.
— Some problems in introducing local government reform in Tanganyika. In: J. of Afr. Adm., Vol. VIII, July 1956.
Lint and Seed Marketing Board, Tanganyika: Report and account (annually).
Little, A. D. (Inc.): Tanganyika industrial development. Dar-es-Salaam, Dec. 1961.
Lock, G. W.: Programme of an agronomic investigation of sisal. Dep. of Agr. Dar-es-Salaam, 1935.
— A study of methods of cultivating sisal in Kenya in comparison with those in Tanganyika. In: E.A.A.J., Vol. II, March 1937.
— Sisal, London, 1962.
Lord, R. F.: Economic aspects of mechanised farming at Nachingwea in the Southern Province of Tanganyika Territory (in print).
Loxton Report: The Kilombero valley land use survey. Dep. of Agr. 1953.
Lunan, M.: Mound cultivation in Ufipa, Tanganyika. In: E.A.A.J., Vol. XVI, Oct. 1951.
— and J. Weir: Maize growing at Ismani, Iringa, Tanganyika. In: E.A.A.J., Vol. XVI, Oct. 1951.
— and D. Brewin: The agriculture of Ukara island. In: Emp. Cot. Grow. Rev., Vol. XXXIII, No. 4, Oct. 1956.
Lussy, P. K.: Die Wapogoro. In: Anthropos, Vol. XLVI, 1951.
— Some aspects of work and recreation among the Wapogoro of Southern Tanganyika. In: Anthropological Quarterly, Vol. XXVI, No. 4, Oct. 1953.
Luytjes, A.: Agrarische Politiek in Brits-Oost-Afrika. In: Tropic. Prod. Dep. of the Royal Trop. Inst. No. 242, Amsterdam, 1955.
Mackenzie, D. R.: The spirit-ridden Konde. Philadelphia, 1925 (p. 109 ff.).
Mackinnon, E.: Report on the possibilities of agricultural development by mechanisation of the Bukindo area of Uzinza Peninsula. C.D.C., London, 1951.
Makwaia, N.: Notes on land, cattle and water in Shinyanga district. n. d. MS.
Malcolm, D. W.: A report on land utilisation in Sukuma. Appendix, 1936, MS.
— Ukara, a report on economic conditions on the island of Ukara with special reference to soil erosion and other matters influencing production. 1934, MS.
— Sukumaland. An African people and their country. Int. Afr. Inst. London, 1953.
Mann, H. H.: Tea cultivation in the Tanganyika Territory and its development. Millbank, 1953.
Markham, A. E. G.: Orientation of food production programmes: Animal production. In: FAO/WHO Seminar 1959 (?).

Marsland, H.: Mlau cultivation of African agriculture. In: Afr. Aff. Jan. 1954 (see also T.N.R. No. 5, 1938).
— Collection and commercial preparation of cerea rubber. Dar-es-Salaam, 1943.
Martin, C. J.: Some estimates of the general age distribution, fertility and rate of natural increase of the African population of British East Africa. In: Population Studies 7, 2. Nov. 1953.
Masefield, G. B.: Agricultural extension methods amongst African peasant farmers. In: E.A.A.J., Vol. XII, April 1946.
Mason, I. L., and J. P. Maule: The indigeneous livestock of Eastern and Southern Africa. Com. Agr. Bur. 1960.
Massell, B. F.: The distribution of economic gains in an East Africa Federation. African Studies Association. Oct. 1963, San Francisco.
— The trade between East Africa and neighbouring countries. In: Economic and Statistical Review No. 4, Sept. 1962.
Matheson, H. K., and E. W. Bovill: East African agriculture. Oxford Univ. Press, 1950.
May, W. B.: Horticultural note on the raising of cocoa in East Africa. E.A.A.F.R.O. Horticultural Technical Note No. 3, 1960, Mimeo.
McGregor, C. J.: Flue-cured tobacco. Dep. of Agr. Pamphlet No. 18, Dar-es-Salaam, 1937.
— Notes on pyrethrum in Iringa. Dep. of Agr. Leaflet No. 11, n. d.
McKinlay, K. S.: Cotton pest control in the Eastern Province, Tanganyika. In: E.A.A.J., Vol. XXII, July 1956.
— Cotton pests in the Eastern Province, Tanganyika Territory. In: Emp. Cot. Grow. Rev., Vol. XXXIII, No. Oct. 1956.
McMaster, D. N.: Change of regional balance in the Bukoba District of Tanganyika. In: Geographical Rev. Vol. L, No. 1, 1960 (see also: T.N.R. No. 56, March 1961).
Meek, C. K.: Land and custom in the colonies. 2nd ed. London, 1949.
Meek, K. O.: Stock reduction in the Mbulu highlands, Tanganyika. In: J. of Afr. Adm., Vol. 5, Oct. 1953.
Meulen, J. G. J. van der: Report for the Government of Tanganyika on rice production. FAO Rep. No. 919, Rome 1958.
Mhaliga, A.: Co-operation in Tanganyika. Pontificia Universitas Gregoriana, Tabora, 1958.
Michelmore, A. P. G.: The international red locust control. In: T.N.R. No. 20, 1945.
Milne, G.: Amani memoirs. A soil reconnaissance journey through parts of Tanganyika Territory. In: Journal of Ecology, Vol. XXXV, No. 1 and 2, 1947.
— Bukoba. High and low fertility on a laterised soil. In: E.A.A.J., Vol. IV, July 1938.
— Soil type and soil management in relation to plantation agriculture in East Usambara. In: E.A.A.J., Vol. III, July 1937.
— G. H. Gethin-Jones, V. A. Beekley, W. S. Martin, G. Griffith, and L. W. Raymond: A provisional soil map of East Africa. Amani Memoirs, London, (?) 1936.
Moffet, J. P.: Handbook of Tanganyika. Dar-es-Salaam, 1958.
— A review of scientific progress in Tanganyika. In: T.N.R. No. 34, Jan. 1953.
Molochan, M. J. B.: Detribalization, Tanganyika. Dar-es-Salaam, 1959.
Morison, C. G. T.: The mbuga soil of Sukumaland. 1954, Mimeo.
— and B. I. Wright: The soil of Sukumaland, Shinyanga. 1951, MS.

MORS, P. O.: Cattle in Buhaya. In: Anthropological Quarterly, Vol. 27, No. 1, Jan. 1954.
MORSTADT, H.: Kaffee-Schädlinge und -krankheiten Afrikas. In: Tropenpflanzer, Okt. 1935, März 1936, Juli 1936, Nov. 1936 und Febr. 1937.
MORTON, J. F.: The cashew's brighter future. In: Economic Botany, Vol. 15, No. 1, Jan. March 1961.
Moshi Native Coffee Board: Annual reports (ending 1954 ?).
MUNGER, E. S.: African coffee in Kilimanjaro. In: Econ. Geogr., Vol. XXVIII, April 1952.
MURRAY, S. S.: Report on tobacco with special reference to the prospects of increased production in Central and East Africa. H.M.S.O. London, 1949.
MUTTER, N. E. S.: Land tenure in Busukuma. Mwanza 1938 (?) MS.
— and M. BIGGER: Cashew. Min. of Agr. Bull. No. 11, Dar-es-Salaam 1961.
MWALUKO, E. P.: Famine relief in the Central Province of Tanganyika, 1961. In: Trop. Agr., Vol. 39, No. 3, July 1962.
NAPIER-BAX, S.: The grazing of the huru huru mbuga system, Shinzanga. 1942, MS.
— A practical policy for tsetse reclamation and field experiments. I and II. In: E.A.A.J., Vol. IX, July 1943.
NAUMANN, F. A., and H. ABEL: Contributions to the geography and ethnography of the Matengo Highlands, 1930—33. In: Dtsch. Geogr. Blätter, 46, 1951.
NORTHCOTE, R. C.: A memorandum on native land tenure. Dar-es-Salaam, 1945.
NORTHWOOD, P. J.: Cashew production in the Southern Province of Tanganyika. In: E.A.A.J., Vol. XXVIII, July 1962.
NOWACK, E.: Geologie und Landwirtschaft in Ostafrika. In: Tropenpflanzer, Dez. 1938.
NOWELL, W.: The first ten years of the Amani research station. In: Emp. Cot. Grow. Rev., Vol. XIV, No. 2, April 1937.
NUTMAN, F. J.: A study of the Mlalo basin, an area forming part of the west Usambara development scheme, 1945, MS.
— Agave fibres. In: Emp. J. of Exp. Agr., Vol. V, No. 17, Jan. 1937.
NYERERE, J. K.: The second scramble. Dar-es-Salaam, 1961.
— "Ujamaa" The basis of African Socialism. Dar-es-Salaam, 1962.
OBI, O. F.: The coconut industry in Tanganyika. UN. Econ. Com. for Africa, May 1961, Mimeo.
OBST, E.: Das abflußlose Rumpfschollenland im nordöstlichen Deutsch-Ostafrika. Teil II, S. 246 ff. Hamburg 1923.
OCKER, W.: Entwicklungsmöglichkeiten für die Landwirtschaft und den Wasserbau im Rahmen des Drei-Jahres-Planes für Tanganyika. Düsseldorf, 1961, Mimeo.
OLDACKER, A. A.: Tribal customary land tenure in Tanganyika. In: T.N.R. No. 47 and 48, 1957.
— Interim report on tribal customary land tenure in Tanganyika. Dar-es-Salaam, 1957, MS.
OLIVER, R.: The missionary factor in East Africa. London 1952.
ORDE-BROWNE, G. ST. J.: Labour conditions in East Africa. H.M.S.O. Col. No. 1946.
OSTERTAG, R. VON: Die Bekämpfung der Tierseuchen in den zur Zeit unter Mandat stehenden deutschen Kolonien. In: Tropenpflanzer, Aug. 1937.
Overseas Food Corporation: Annual report 1947/48. Scientific Dep. Part. I and II, Mimeo.
— Report of the Kongwa working party. London, 1950.
— Report and accounts, 1948 to 1954—55. H.M.S.O. London.
— Conclusions and recommendations on agricultural development policies in Southern Province, 1950 (?), Mimeo.

Oxley, T. A.: Grain Storage in East and Central Africa. H.M.S.O. London, 1950.
Page-Jones, F. H.: Water in Masai Land. In: T.N.R., Vol. 26, 1948.
Paterson, R. L.: Ukara island. In: T.N.R. No. 44, 1956.
Paulus, M.: Die Aufbauprobleme der Genossenschaften in Tanganyika. Diplomarbeit, Universität Köln, Seminar für Sozialpolitik, 1962.
Peacock, A. T., and D. G. M. Dosser: The gross domestic product of Tanganyika, 1954—57. Dar-es-Salaam, 1959.
— The national income of Tanganyika, 1952—54, Col. Res. Stud. No. 26.
Peat, J. E.: Tanganyika cotton production. In: Emp. Cot. Grow. Rev., Vol. XXIV, No. 3, July 1947.
— Cotton in Tanganyika during the first five years. In: Emp. Cot. Grow. Rev., Vol. XXIII, No. 1, April 1946.
— Fertilizer responses-Ukiriguru. In: Proceedings of a Meeting, Muguga, 1953, Mimeo.
— A note on cotton growing in Tanganyika. In: Emp. Cot. Grow. Rev., Vol. XXX, No. 3, July 1953.
— Cotton production prospects for Tanganyika Territory by 1961 and thereafter. In: Prospective trends in cotton yields and acreage, Washington, 1961.
— and K. J. Brown: A report of cotton breeding for the Lake Province of Tanganyika, seasons 1939—40 to 1957—58. In: Emp. J. of Exp. Agr., Vol. XXIX, No. 114, April 1961.
— Effect of management on increasing crop yields in the Lake Province of Tanganyika. In: E.A.A.J., Vol. XXVI, Oct. 1960.
— The yield responses of rain-grown cotton at Ukiriguru in the Lake Province, Tanganyika, I. The use of organic manure, inorganic fertilizers and cottonseed ash. In: Emp. J. of Exp. Agr., Vol. XXX, No. 119, July 1962.
— and A. N. Prentice: The maintenance of soil productivity in Sukumaland and adjacent areas. In: E.A.A.J., Vol. XV, July 1949.
Phillips, J. F. V.: The nature, setting and agricultural possibilities of the Kongwa region, O.F.C. 1950, Mimeo.
— Report on livestock development possibilities, Kongwa, Tanganyika, 1950, Mimeo.
— Brief report prepared for the information of the committee appointed by the board on the recommendations and findings of the Southern Province Land Utilization Committee. 1950, Mimeo.
Pike, A. H.: Soil conservation amongst the Matengo tribe. In: T.N.R. No. 6, 1938.
Pole Evans, I. B.: The vegetation of East and Central Africa. Pretoria, 1948.
Popplewell, G. D.: Notes on the Fipa. In: T.N.R. No. 2, 1937.
Pritwitz, G. von: Die Oberflächengestaltung der Gebirgslandschaft Utschungwe im östlichen Mittelafrika. Berlin 1932.
Prentice, A. N.: Tie-ridging with special reference to semi-arid areas. In: E.A.A.J., Vol. XII, Oct. 1946.
— The cotton belt of East Africa. In: Emp. Cot. Grow. Rev., Vol. XXXI, No. 4, Oct. 1954.
Purvis, J. T.: Agricultural history of Musoma district. Musoma, 1945, MS.
— The Sukumaland development scheme. In: Corona, Febr. 1951, Vol. III, No. 2.
— B. J. Dudbridge, and J. B. Clegg: The mechanisation and capitalisation of peasant agriculture. (Sukumaland) Malya, 1952, MS.
Pyrethrum Post: Official publication of the African pyrethrum research council.
Reaburn, R. J.: Some economic aspects of African agriculture. In: E. A. Econ. Rev. Jan. 1959.
Ratcliffe, A.: Über Kokospalmenkultur. In: Tropenpflanzer, April, 1920.

RATSEY, M.: Animal husbandry on Ukara island and effects on the diet of the inhabitants. Mwanza, 1956, MS.

RAYMOND, W. D.: Nutrition and Tanganyika. In: T.N.R. No. 11, April, 1941.

REDFEARN, C., and N. R. FUGGLES-COUCHMAN: Large-scale wheat production at Oldeani, Tang. Terr., in 1943 and 1944. In: E.A.A.J., Vol. XI, Oct. 1945.

REDMAYNE, A.: Preliminary report on a Hehe community. In: East Afr. Inst. of Soc. Res. Conf. 1962, Mimeo.

REICHART, K.: Die Besiedlungsfähigkeit der ostafrikanischen Bruchstufe vom Marang-Urwald bis zum Ndessekera und der westlichen angrenzenden Gebiete. In: Tropenpflanzer, Okt. 1928.

REINING, P. C.: Village organization in Buhaya. In: East Afr. Inst. Soc. Res. Conference 1952, Mimeo (?).

— Haya Land Tenure: landholdings and tenancy. In: Anthropological Quarterly, Vol. 35, No. 2, April 1962.

RENSBURG, H. J. VAN: Conservation of vegetation in Tanganyika Territory. In: Management and conservation of vegetation in Africa, Bull. No. 41, Penglais, Wales 1951.

— Land usage in semi-arid parts of Tanganyika. In: E.A.A.J., Vol. XX, April 1955.

— Notes of fodder and pasture grasses in Tang. Terr. In: E.A.A.J., Vol. XIII, Jan. 1948.

— Pasture management. In: E.A.A.J., Vol. XIV, March 1949.

— The role of pasture development in soil conservation, Tang. Terr. In: E.A.A.J., Vol. XIII, July 1947.

Report of an informal committee under the chairmanship of Lord Haily: Native land tenure in Africa, C. M. No. 10, London, 1945.

Report of the Conference on African Land Tenure in East and Central Africa: In: J. of Afr. Adm. Oct. 1956.

Report of the United Nation visiting Mission to Tanganyika, 1957. Dar-es-Salaam, 1958.

Report on the culture of cocoa in the Tanga Province, Tanganyika, 1960 (?), Mimeo.

Report on the establishment of a cane sugar industry on the Great Ruaha River in the Kilombero valley in Tanganyika. 1960 (?), Mimeo.

Report to the Council of the League of Nations on Tanganyika Territory, 1937.

RICHARDS, A.: East African chiefs, London, 1959.

RICHARDSON, R. B.: Peasant holdings in the Western Province and the certificate used at Kingolwira, n. d. MS.

RIGBY, P. J. A.: Aspects of residence and co-operation in a Gogo village. In: E. A. Institute of Social Research, Kampala, Conf. Jan. 1962, Mimeo.

RINGER, K.: Die Agrarverfassungsprobleme Tanganyikas in ihrem Zusammenhang mit den Fragen und Aufgaben der volkswirtschaftlichen Entwicklung des Landes. Diss. Freiburg, 1950.

ROBERTSON, B.: Report regarding the proposed settlement of Indian agriculturists in Tanganyika Territory. Cmd. 1312, 1921.

ROBERTSON, J. K.: Mixed or multiple cropping in native agricultural practice. In: E.A.A.J., Vol. VI, 1941.

— and C. W. ROMULOW-PEARSE: A system of cultivation in the Mbeya District of Tanganyika. In: E.A.A.J., Vol. XX, Jan. 1955.

— The growing of canning beans in Tanganyika. In: World Crops. Vol. 7, No. 1, Jan. 1955.

ROBINSON, J. B. D.: Arabica coffee production in Tanganyika. In: T.C.N., March 1962.

— Notes on coffee growing in the Oldeani area. In: T.C.N., April 1962.

Rohde, H.: Einiges über Kaffeeanbau am Meru mit künstlicher Bewässerung. In: Tropenpflanzer, Juli/Aug. 1924.

Roth, W. J.: Co-operatives and the economic aspects of cultural change among the Basukuma of Tanganyika. Diss. Cath. Univ. Washington (in preparation).

Rounce, N. V., and G. Milne: The unsuitability of certain virgin soils to the growth of grain crops. In: E.A.A.J., Vol. I, No. 2, 1936.

— Agricultural development in relation to land use in Sukumaland. In: Proceedings of the UN, Scient. Conf. on the Cons. and Util. of Res., New York, 1948.

— The agriculture of the cultivation steppe of the Lake, Western and Central Province of Tanganyika. Cap Town, 1949.

— The development, expansion and rehabilitation of Sukumaland. In: Emp. Cot. Grow. Rev. Jan. 1949.

— The economic development of Sukumaland. In: Emp. Journ. of Exp. Agr. No. 19, 1951.

— Memorandum on the mechanised cultivation schemes in Sukumaland. Mwanza, 1951, MS.

— Individual native smallholdings. In: E.A.A.J., Vol. III, 1937.

— Soil classification and survey in relation to land rehabilitation and expansion in Usukuma. In: Proceed. of Conf. on Trop. and Sub-Trop. Soils, 1948. Harpenden, 1948.

— Survey of crop acreages in Sukumaland, 1945, Mimeo.

— Survey of crop yields in Sukumaland, 1949, Mimeo.

— Technical considerations on the economic development of Sukumaland. In: Emp. J. of Exp. Agr., Vol. XIX, No. 76, Oct. 1951.

— and D. Thornton: The ridge in native cultivation, Mwanza District. In: E.A.A.J., Vol. IV, March 1939.

— J. G. M. King, and D. Thornton: A record of investigations and observations on the agriculture of the cultivation steppe of Sukuma and Nyamwezi with suggestions as to the lines of progress. Dar-es-Salaam, 1942.

Rungwe African Co-operative Union: Annual reports, Tukuyu, (Coffee).

Russel, E. W. ed.: The natural resources of Eest Africa. Nairobi, 1962.

Ruthenberg, H.: Ansatzpunkte zur landwirtschaftlichen Entwicklung in Tanganyika. Berlin 1962.

Savile, A. H.: A study of recent alterations in the flood regions of three important rivers in Tanganyika. In: E.A.A.J., Vol. XI, Oct. 1945.

Saers, G. F. ed.: Handbook of Tanganyika. London, 1930.

Schaegelein, T.: La tribu des Wagogo. In: Anthropos. Vol. XXXIII, 1938.

Scheel, J. O. W.: Tanganyika und Sansibar. Bonn, 1959.

— Das Genossenschaftswesen in der Wirtschaft der Afrikaner in Tanganyika. In: Afrika-Informationsdienst. Jg. 3, Nr. 6, 1960.

Scheerder, R. P., and R. Tastevin: Les Waluguru. In: Anthropos, Vol. 45, June 1950.

Scherer, J. H.: The Ha of Tanganyika. In: Anthropos, Vol. 54, 1959.

Scheven, A.: Chibuga of Ukara. 1959, MS.

Scott, R. R.: Preliminary survey of the position in regard to nutrition amongst the native of Tanganyika Territory. Dar-es-Salaam, 1937.

Seabrook, A. T. P.: The groundnut scheme in retrospect. In: T.N.R. No. 47—48, 1957.

Senior, H. S.: The Sukuma homestead. In: T.N.R. No. 9, 1940.

Shapestone, D.: The oil palm in Western Tanganyika. In: E.A.A.J., Vol. XVI, Jan. 1951.

SHAW, J. V.: The development of African local government in Sukumaland. In: J. of Afr. Adm., Vol. VI, No. 4, 1951.

SIBLEY-WARNE, W.: The vegetable garden on the coast. Dep. of Agr. Leaflet, No. 10, 1937.

SIEMER, H. D.: Das Genossenschaftswesen in Tanganyika und seine Aufgaben in der zukünftigen Entwicklung des Landes. Berlin 1963, MS.

SKEFFINGTON, A.: Tanganyika in transition. Fabian Research Series, 212. London 1960.

SMITH, E. H. G.: Papain: Its production and market. H.M.S.O. London, 1953.

SMITH, E. W.: The earliest ox-wagon in Tanganyika. An experiment which failed. In: T.N.R. No. 40, 1955.

SMITH, H. C.: The Sukuma system of grazing rights. In: E.A.A.J., Vol. IV, 1938.

SPRY, J. F.: Some notes on land tenure, adjudication of rights and registration of titles with specific reference to Tanganyika. In: J. of Afr. Adm., Vol. VIII, Oct. 1956.

STAHL, K. M.: Tanganyika, Soil in the Wilderness. s'Gravenhage, 1961.

STANNER, W. E. H.: Sociological problems of the groundnut scheme in Tanganyika. In: Col. Rev., Vol. 6, No. 2.

STAPLES, R. R.: Pasture investigations in Tanganyika. In: Trop. Agr., Vol. IX, No. 12, Dec. 1932.

— Notes on water conservation in sub-arid Tanganyika. In: Dept. Vet. Sci. Anim. Husb. An. Rep. 1933.

— Conservation of water supplies in semi-arid East Africa (Mpwapwa). In: Vet. Dep. T. T. An. Rep. 1937.

— Combating soil erosion in the Central Province of Tang. Terr. In: E.A.A.J., Vol. VII, Oct. 1941.

STENHOUSE, A. S.: Matengo pit system of cultivation. In: E.A.A.J., Vol. X, July 1944.

STOCKDALE, F.: Report on visit to East Africa. Colonial Advisory Council of Agriculture and Animal Health. Colonial Office, 1937.

STOCKLEY, C. M.: Phosphate deposits in Tang. Terr. In: E.A.A.J., Vol. XII, Oct. 1946.

STOREY, H. H.: Basic research in agriculture, a brief history of research at Amani, 1928—1947. East Afr. Agr. and Forest Res. Org., Nairobi, n. d.

STURDY, D., W. E. CALTON, and G. MILNE: Survey of waters of Mount Meru. In: J. East Afr. Uganda Nat. Hist. Soc. No. 45, 1933.

— Agricultural notes (Wambulu, Mbulu District) In: T.N.R. No. 1, 1936.

SWYNNERTON, C. F. M.: An experiment in control of tsetse fly at Shinyanga, Tanganyika. Imp. Bu. of Ent. 1925.

— The tsetse flies of East Africa. London 1936.

SWYNNERTON, R. J. M.: The improvement of the coconut industry on the Tanga Coast. In: E.A.A.J., Vol. XII, Oct. 1946.

— Onion cultivation on Kilimanjaro. In: E.A.A.J., Vol. XII, Jan. 1947.

— Some problems of the Chagga on Kilimanjaro. In: E.A.A.J., Vol. XIV, Oct. 1949.

— and A. L. B. BENNET: All about KNCU coffee. Moshi, 1948.

Tanganyika: Agricultural credit agency. Dep. of Agr. 1961, Mimeo.

— Agriculture in Tanganyika. Dep. of Agr., 1945.

— An agronomic history of Songea coffee. Dep. of Agr. 1956, MS.

— An agronomic history of Songea fire-cured tabacco. Dep. of Agr. 1956, MS.

— Annual report of the Board of the Land Bank of Tanganyika.

— Annual report on co-operative development.

— Annual report of the Department of Agriculture. Parts I. and II.

Tanganyika: Annual report of the Department of Commerce and Industry.
— Annual report of the Department of Lands and Surveys.
— Annual report of the Department of Veterinary Services.
— Annual report of the Department of Water Development and Irrigation.
— Annual report on Development, 1949—50.
— Annual report of the Grain Storage Department, 1952—53 to 1956—57.
— Annual report of the Tsetse Survey and Reclamation Department.
— Annual report. Tropical Pesticides Institute, Arusha.
— Atlas of Tanganyika. Department of Lands and Surveys, 3d ed., Dar-es-Salaam, 1956.
— Budget survey (annually).
— Arusha Development Plan, Reconnaissance report of the area Ardai-Lolkisale-Makuyuni, 1958 (?), Mimeo.
— A career in agriculture. Dep. of Agr., Dar-es-Salaam, 1960.
— Central development committee. 1. Draft report 1939. 2. Memoranda 1—25. 3. Minutes, Meetings 1—15 and 16—25. 4. Report, Dar-es-Salaam, 1940. 5. Report of the provincial committees part 1 and 2, Dar-es-Salaam, 1942, Mimeo.
— Coffee marketing policy. Legislative Council. Gov. Pap. No. 3, 1959.
— Comments on the report by Dr. Dolfi on "Nyumba ya Mungu Storage Scheme". Dep. of Water Dev. and Irrig., Nov. 1960, Mimeo.
— Committee of enquiry to examine the basis of ginners costs and renumeration. Dar-es-Salaam, 1953.
— Commodity marketing and price stabilisation, 1962, Mimeo.
— Community development policy. Minist. of Co-operative and Community Development, 1962, Mimeo.
— The cotton industry, 1939—53. Dar-es-Salaam, 1953.
— The cotton industry, questions and answers. Dar-es-Salaam, 1954.
— De-stocking in the Lake Province. Dar-es-Salaam, 1956.
— Development plan 1955—60 (capital works programme). Dar-es-Salaam, 1955.
— The development plan for the Lindi District, 1954—1960.
— Development Plan for Tanganyika 1961/62—63/64. Dar-es-Salaam, 1961.
— Draft proposals for co-operative copra drying kiln. Dar-es-Salaam District. Dep. of Agr. n. d. MS.
— Employment and earnings in Tanganyika, 1961. Dar-es-Salaam, 1962.
— Ex-German tea estates. Dep. of Lands and Waters. Dar-es-Salaam, 1949.
— A guide for primary produce-marketing co-operative societies. Dep. of Co-op. Dev. Moshi, 1951.
— Household rural surveys in Tanganyika. Pilot survey in Bagamoyo-June 1961. Dar-es-Salaam. Mimeo.
— Land development survey. 1. Report: 1928/29 Iringa Province. 2. Report: Iringa Province 1930. 3. Report: Uluguru Mountains, 1929/30. 4. Report: Mbulu District, 1930. 5. Report: Eastern Prov. 1932. (Bagshawe, F. J., H. Wolfe, C. J. McGregor, M. F. Bell, W. J. Hill).
— Local development and African productivity loan funds. Report and Accounts (annually).
— Meeting of the East African regional committee for the conservation and utilization of soil. Dep. of Agr. Dodoma, 1956, Mimeo.
— Memorandum on closing land in Tanganyika Territory. Dar-es-Salaam, 1937.
— The Meru land problem. Legislative Council, 1952.
— Monthly statistical bulletin. Econ. and Stat. Div. of the Treasury.
— An outline of post-war-development proposals. Dar-es-Salaam, 1944.
— Pack donkey transport. Vet. Dep. Dar-es-Salaam, 1938.

— Pilot survey of African agriculture. Dar-es-Salaam, 1960 (?), Mimeo.
— A plan for sampling surveys of African agriculture. East African Statistical Department, 1954. Mimeo.
— Proposals of the Tanganyika Government for land reform. Gov. Pap. No. 2, Dar-es-Salaam, 1962.
— Record of research work carried out. Dep. of Agr. Mimeo (annually).
— Reorganization of the Ministry of Agiculture, 1st. July 1962, Mimeo.
— Report on the African loan funds (annually).
— Report of the Commission appointed to consider and advise on questions relating to the supply and welfare of native labour in the Tanganyika Territory.
— Report to the enquiry on coffee export tax. 1955.
— Report of a committee of enquiry appointed to investigate the dairy and livestock industry in Northern Province. Dar-es-Salaam, 1944.
— Report of a committee appointed by His Excellency the Governor to submit proposals in connection with land development and the provision af financial assistance to settlers and planters. Sess. Pap. No. 3, 1930.
— Report of the committee on manpower. Dar-es-Salaam, 1951.
— Report of the development of Mbulu District, Dar-es-Salaam, 1951.
— Report on the Ikowa irrigation scheme, Central Province, Dar-es-Salaam, 1955, Mimeo.
— Report on an inquiry into agricultural education at primary and middle schools. Dep. of Agr., Dar-es-Salaam, 1956.
— Report on the Lake Province agricultural show and trade exhibition, Mwanza, 1933, Mimeo (see also: Dar-es-Salaam, Exhibition 1929).
— Report of land usage survey Madege-Mbulu, April 1961, MS.
— Report of Mlado rehabilitation scheme. Dep. of Agr. n. d. Mimeo.
— Report on Muheza citrus development scheme, Tanga Province. Dep. of Agr. n. d. Mimeo.
— Report on the Pyrethrum Board 1949/52. Dar-es-Salaam, 1955.
— Report on the state of industrial relations in the sisal industry. Dar-es-Salaam, 1959.
— Report on survey of inter-territorial trade between Tanganyika and Northern Rhodesia and Nyasaland, Dar-es-Salaam, 1954.
— A review of development plans in the Southern Province, 1953. Dar-es-Salaam, 1954.
— Review of land tenure policy. Gov. Pap. No. 6, Dar-es-Salaam, 1958.
— Revised Development and Welfare Plan for Tanganyika, 1951.
— Role of livestock in the substistence and cash economy of Tanganyika. Vet. Dep. 1962, Mimeo.
— Specimen tasks for labour employed in agricultural and industrial undertakings. Labour Dep. Dar-es-Salaam, 1950.
— Statistical abstracts. Dar-es-Salaam (annually).
— Sugar policy. Gov. Pap. No. 9, 1958, Gov. Pap. No. 1, 1959.
— Survey of industrial production 1958. E.A.H.C. State Dep. Tang. Unit. 1960.
— A Ten-Year Development and Welfare Plan for Tanganyika Territory, 1946.
— Text book of agricultural practices: for use in primary and middle schools, Tanga Province, 1958.
— Trade Report (annually).
— Two pilot schemes to solve the overstocking problem in Tanganyika. Dep. of Vet. Serv. In: Trop. Agr., Vol. 32, No. 1, Jan. 1955.
— Wheat and wheat flour policy. Legislative Council. Gov. Pap. No. 8, 1958.

Tanganyika Agricultural Corporation: Reports and Accounts (annually).
Tanganyika Coffee Board: A handbook of arabica coffee in Tanganyika. Moshi, 1959, Mimeo.
— Research reports, Lyamungu (annually).
Tanganyika Coffee Growers Association, Ltd.: Annual report, Moshi.
Tanganyika Co-operative Trading Agency Ltd.: Annual report, Moshi.
Tanganyika and Kenya Sisal Growers Association: Report of the Joint Marketing Committee, Tanga, 1948.
Tanganyika National Assembly: Assembly Debates: official report (ending 1961).
Tanganyika Parliamentary Debates: Official report (beginning 1962).
The Tanganyika Planting Company: Nairobi 1961 (?) (Arusha-Chini).
Tanganyika Sisal Growers Association: Sisal research station. Annual reports, see also: Supplements to annual reports.
Tanganyika: Trade Bulletin (Ministry of Commerce).
— A revised policy for the development and rehabilitation of Sukumaland, Mimeo o. J. (1949 ?).
— Co-operative movement in the cotton growing areas of the Lake Province. Mimeo, 1962 (?).
— Agricultural credit with particular reference to pilot scheme for credit through marketing in the co-operatives in the Lake Region, Tanganyika, Mimeo, 1963 (?).
Tanganyika Agricultural Corporation: Pasture research station, Kongwa. Record of research work, 1959—60, Mimeo.
Tanganyika: Agricultural report on certain areas of Iringa Province: Dabaga cloud forest, Ukwama, Kisinga, MS.
TANNER, R. E. S.: Law enforcement by communal action in Sukumaland. In: J. of Afr. Adm., 1955, S. 159.
— A preliminary enquiry into Sukuma diet in the Lake Province. Tanganyika Territory. In: E. Afr. Med. J. No. 3, 1956.
— Land tenure in Northern Sukumaland. In: E.A.A.J., Vol. XXI, Oct. 1955.
— Small holdings in the Tanga Coast. In: E.A.A.J., Vol. XXIV, Oct. 1958.
— Population changes 1955—59 in Musoma District, Tanganyika, and their effect on land usage. In: E.A.A.J., Vol. XXVI, Jan. 1961.
TAPLEY, R. G.: Copper spraying doubles coffee yields in small holdings. In: T.C.N., July 1962.
TAUBER, I. B.: The population of Tanganyika. United Nations, 1949.
TAYLOR, J. C.: The political development of Tanganyika. Stanford and London 1963.
Tea Research Institute of East Africa: Quarterly circular and annual reports.
TEALE, E. O.: The soil and agricultural development in relation to the geology of portions of the Northern Kigoma and Southern Bukoba Province, Dar-es-Salaam, 1929.
— and C. GILLMAN: Report on the investigation of the proper control of water and reorganization of the water boards in the Northern Province of Tanganyika. Dar-es-Salaam, 1935.
— Undeveloped land in East Africa with special reference to Tanganyika. In: Emp. Cot. Grow. Rev., Vol. XXV, No. 1, Jan. 1948.
TELFORD, A. M.: Report on the development of the Rufiji and Kilombero Valleys. Gov. of Tanganyika, London 1929.
TEW, M.: The peoples of the Lake Nyasa Region. Oxf. Un. Press, London 1951.
The Sisal Review: (Industrial Fibres Review).

THORNTON, D., and N. V. ROUNCE: Ukara island and the agricultural practices of the Wakara. In: T.N.R., No. 1, 1936.

THURSTON, J. L.: Human resources and manpower planning in Tanganyika. Ford Foundation, 1962. Mimeo.

TIPPER, L. H. C.: Report on tour of cotton-ginning industries in the Sudan, Uganda and Tanganyika. In: Emp. Cot. Grow. Rev., Vol. XXXVII, July 1960.

TREITZ, W.: Ostafrikanische Entwicklungsgebiete. Bundesstelle für Außenhandelsinformationen. Köln 1961.

TROLL, C., and K. WIEN: Oldeani — Ngorongoro, wiss. Veröff. des Museums zu Leipzig, Leipzig 1935.

TUCKETT, J. R.: Pyrethrum. Min. of Agr. Bull. No. 5, Dar-es-Salaam, 1961.

THWAITES, H.: Wanyakyusa Agriculture. In: E.A.A.J., Vol. IX, April 1944.

United Nations: Ninth report of the committee on rural economic development of the trust territories. June 1960, Mimeo.

URQUHART, D. H.: Report on prospects for cocoa in Tanganyika, 1959 MS.

WAKEFIELD, A. J.: Peasant holdings, address given to the Brit. Med. Ass., Dar-es-Salaam, 1936, Mimeo.

— The groundnut scheme. In: E.A.A.J., Vol. XIII, Jan. 1948.

— Mixed farming and peasant holding in Tanganyika Territory. In: Emp. Cot. Grow. Rev., April 1934, No. 2, Vol. XI.

— Native production of coffee on Kilimanjaro. In: Emp. Jour. of Exp. Agr., Vol. VI, No. 14, April 1936.

— D. L. MARTIN, and J. ROSA: Report of a Mission to investigate the practicability of the mass production of groundnuts in East and Central Africa. Appendix, Tang. Terr. Reprint. Dar-es-Salaam, 1947 (?).

WALDRON, C. A.: Expansion of coffee, cocoa and oil palm (Kilombero). Dep. of Agr. 1958, Mimeo.

— Kilombero Valley rubber survey. Dep. of Agr. 1960, Mimeo.

WALLACE, G. B.: The establishment and running of a papaw plantation. Tanganyika. In: E.A.A.J., Vol. XIII, April 1948.

— and M. M. WALLACE: List of plant diseases of economic importance in Tanganyika. Comm. Mycological Inst. Pap. 26.

WAZIRI, JUMA: The Sukuma societies for young men and women. In: T.N.R. No. 54, 1960 (see also No. 5).

WEIGT, E.: Europäer in Ostafrika. Köln 1955.

WEITZENBERG, H.: Der Bukobabezirk in Deutschostafrika. In: Tropenpflanzer, Juni 1942, Juli/Aug. 1942 und Sept. 1942.

WIGGLESWORTH, A.: The future of the sisal industry in East Africa, 1946.

WILBRANDT, H., und H. RUTHENBERG: Bericht über Tanganyika, Sachgebiet Landwirtschaft, Berlin 1961, Mimeo.

WILLIAMS, O. G.: Village organization among the Sukuma. In: Man. Vol. XXXV, 1935.

WILSON, G.: The land rights of individuals among the Nyakyusa. In: Rhodes-Livingstone, Paper No. 1, 1938.

WILSON, J. M.: Report on the Arusha-Moshi land commission. Dar-es-Salaam, 1947.

WILSON, M.: Good company (Nyakusa) Oxf. Un. Press 1951.

— The Nyakusa. In: Seven Tribes of British Central Africa. Ed. E. Colson, 1951.

WINTER, E. H.: Livestock Markets among the Iraqw of Northern Tanganyika. In: Markets in Africa. Northwestern Univ. Press. 1962.

— Some aspects of political organization and land tenure among the Iraqw. Arusha, 1955, Mimeo.

WOOD, A.: The groundnut affair. London, 1950.

Woods, W.: Report on a fiscal survey of Kenya, Uganda and Tanganyika. Nairobi 1946.

Worthington, E. B.: A survey of research and scientific services in East Africa. 1947—1956, Nairobi.

Wright, A. C. A.: Part I, Land tenure in Busukuma; Part II, Cattle holding and animal husbandry in Central Busukuma; Appendix: Application of mechanical cultivation methods within Sukuma society. Mimeo, 1952 (?).

— Sociology in Sukumaland. In: Corona, March, 1953, Vol. V, No. 3.

Wright, F. C.: African consumers in Nyasaland and Tanganyika. Col. Res. Stud. No. 17, London, 1955.

Wunder, B.: Baumwollbau und -züchtung in Deutsch-Ostafrika. In: Tropenpflanzer, Nr. 4, Aug./Dez. 1923.

Young, R., and H. A. Fosbrooke: Land and politics among the Luguru of Tanganyika. London, 1960.

— Smoke in the Hills: Political Tensions in the Morogoro District of Tanganyika. Northwestern University Press, 1960.

Zimmermann, A.: Rückblick auf die Tätigkeit des biologisch-landwirtschaftlichen Instituts Amani. In: Tropenpflanzer, Sept./Okt. 1924.

ldditional information of this book

gricultural Development in Tanganyika;978-3-540-03088-1) is provided:

p://Extras.Springer.com